COVID-19 and Social Sciences

COVID-19 and Social Sciences

Editors

Carlos Miguel Ferreira
Sandro Serpa

MDPI • Basel • Beijing • Wuhan • Barcelona • Belgrade • Manchester • Tokyo • Cluj • Tianjin

Editors
Carlos Miguel Ferreira
Interdisciplinary Centre of Social
Sciences—CICS.NOVA
Portugal

Sandro Serpa
University of The Azores, Portugal;
Interdisciplinary Centre of Social
Sciences—CICS.UAc/CICS.NOVA.UAc;
Interdisciplinary Centre for Childhood
and Adolescence—NICA—UAc

Editorial Office
MDPI
St. Alban-Anlage 66
4052 Basel, Switzerland

This is a reprint of articles from the Special Issue published online in the open access journal *Societies* (ISSN 2075-4698) (available at: https://www.mdpi.com/journal/societies/special_issues/COVID-19_social_sciences).

For citation purposes, cite each article independently as indicated on the article page online and as indicated below:

LastName, A.A.; LastName, B.B.; LastName, C.C. Article Title. *Journal Name* **Year**, *Volume Number*, Page Range.

ISBN 978-3-0365-0154-3 (Hbk)
ISBN 978-3-0365-0155-0 (PDF)

© 2021 by the authors. Articles in this book are Open Access and distributed under the Creative Commons Attribution (CC BY) license, which allows users to download, copy and build upon published articles, as long as the author and publisher are properly credited, which ensures maximum dissemination and a wider impact of our publications.

The book as a whole is distributed by MDPI under the terms and conditions of the Creative Commons license CC BY-NC-ND.

Contents

About the Editors vii

Carlos Miguel Ferreira and Sandro Serpa
COVID-19 and Social Sciences
Reprinted from: *Societies* **2020**, *10*, 100, doi:10.3390/soc10040100 1

Abraham Rudnick
Social, Psychological, and Philosophical Reflections on Pandemics and Beyond
Reprinted from: *Societies* **2020**, *10*, 42, doi:10.3390/soc10020042 5

Manuel Arias-Maldonado
COVID-19 as a Global Risk: Confronting the Ambivalences of a Socionatural Threat
Reprinted from: *Societies* **2020**, *10*, 92, doi:10.3390/soc10040092 9

Carlos Miguel Ferreira, Maria José Sá, José Garrucho Martins and Sandro Serpa
The COVID-19 Contagion–Pandemic Dyad: A View from Social Sciences
Reprinted from: *Societies* **2020**, *10*, 77, doi:10.3390/soc10040077 27

Catherine Tobin, Georgia Mavrommati and Juanita Urban-Rich
Responding to Social Distancing in Conducting Stakeholder Workshops in COVID-19 Era
Reprinted from: *Societies* **2020**, *10*, 98, doi:10.3390/soc10040098 47

Dan Grabowski, Julie Meldgaard and Morten Hulvej Rod
Altered Self-Observations, Unclear Risk Perceptions and Changes in Relational Everyday Life: A Qualitative Study of Psychosocial Life with Diabetes during the COVID-19 Lockdown
Reprinted from: *Societies* **2020**, *10*, 63, doi:10.3390/soc10030063 61

Fernando Ferri, Patrizia Grifoni and Tiziana Guzzo
Online Learning and Emergency Remote Teaching: Opportunities and Challenges in Emergency Situations
Reprinted from: *Societies* **2020**, *10*, 86, doi:10.3390/soc10040086 75

About the Editors

Carlos Miguel Ferreira holds a Ph.D. in Sociology from UNL—Nova University of Lisbon, and a Master's in Sociology from the same institution. He is an Invited Assistant Professor at the Estoril Higher Institute for Tourism and Hotel Studies. He is a member of the Interdisciplinary Centre of Social Sciences—CICS.NOVA. He is the founder of the Mediterranean Institute (1989) of FCSH-UNL (Faculty of Social Sciences and Humanities of the Nova University of Lisbon); the founder of the Institute of Sociology and Ethnology of Religions (1989) of FCSH-UNL; and the founder of the Institute for Studies and Sociological Divulgation (1992) of FCSH-UNL. He carries out research in the following areas: sociology of health and illness, in particular, the medicalization process; research methodology; sociology of organizations and general sociology. He is the author and co-author of numerous scientific publications and presentations on these topics.

Sandro Serpa has been a higher education faculty member since 2000. He is an Assistant Professor in the Department of Sociology of the Faculty of Social Sciences and Humanities of the University of the Azores since 2013. He received his Ph.D. in Education in 2013 from the University of the Azores, specializing in Sociology of Education. He is an integrated researcher at the Interdisciplinary Centre of Social Sciences—Campus of the University of the Azores, CICS.NOVA.UAc.

He has teaching experience in areas such as research, sociology of education, introduction to sociology, general sociology, sociology of organizations, psychosociology of educational organizations and human resources, among others.

He has more than 230 publications in international journals, books, and other scientific outlets in Brazil, Canada, Germany, India, Kazakhstan, Netherlands, Pakistan, Poland, Portugal, Romania, Switzerland, Turkey, United Kingdom, and United States of America.

His research interests are in teaching of sociology in higher education; sociology of education; sociology of organizations; organizational culture of educational organizations; and scientific communication.

Prof. Dr. Sandro Serpa is Professor in the Department of Sociology at the Faculty of Social and Human Sciences, University of The Azores, Portugal. He is also affiliated with the Interdisciplinary Centre of Social Sciences—CICS.UAc/ CICS.NOVA.UAc, Portugal, and Interdisciplinary Centre for Childhood and Adolescence—NICA—UAc, Portugal.

Editorial

COVID-19 and Social Sciences

Carlos Miguel Ferreira [1,2] and Sandro Serpa [3,4,5,*]

1 Interdisciplinary Centre of Social Sciences—CICS.NOVA, 1649-026 Lisbon, Portugal; carlos.ferreira@eshte.pt
2 Estoril Higher Institute for Tourism and Hotel Studies, 1069-061 Lisbon, Portugal
3 Department of Sociology, Faculty of Social and Human Sciences, University of the Azores, 9500-321 Ponta Delgada, Portugal
4 Interdisciplinary Centre of Social Sciences—CICS.UAc/CICS.NOVA.UAc, University of the Azores, 9500-321 Ponta Delgada, Portugal
5 Interdisciplinary Centre for Childhood and Adolescence—NICA—UAc, University of the Azores, 9500-321 Ponta Delgada, Portugal
* Correspondence: sandro.nf.serpa@uac.pt

Received: 14 December 2020; Accepted: 14 December 2020; Published: 16 December 2020

The COVID-19 pandemic (caused by the Severe Acute Respiratory Syndrome Coronavirus 2, SARS-CoV-2) is having profound effects on all dimensions of life, such as the individual, social, cultural, public health, and economic dimensions [1,2]. However, the place ascribed to social sciences and their contributions is not sufficiently valued, as may be seen in the bibliographic study by Aristovnik, Ravšelj, and Umek [3] (pp. 1 and 22) in an extensive yet very current and illuminating citation:

The empirical results indicate the domination of health sciences in terms of number of relevant publications and total citations, while physical sciences, social sciences, and humanities lag behind significantly. Nevertheless, there is evidence of COVID-19 research collaborations within and between different subject area classifications, with a gradual increase in the importance of non-health scientific disciplines. The findings emphasize the great need for a comprehensive and in-depth approach that considers various scientific disciplines in COVID-19 research so as to benefit not only the scientific community but evidence-based policymaking as part of efforts to properly respond to the COVID-19 pandemic. [. . .] In order to address the economic, socio-cultural, political, environmental, and other (non-medical) consequences of the COVID-19 pandemic, in the near future, COVID-19 must appear higher up on the research agenda of non-health sciences, particularly social sciences and humanities.

This relevance of the relationship between COVID-19 and social sciences results from the fact that the disease is shaped by cultural elements that are studied in social sciences. Ferreira and Serpa [4] maintain, in this regard, that epidemics and pandemics have varied effects on societies. These effects are noticeable at the level of societies' beliefs, institutions, and social, demographic, economic, and political structures. Regarding the notion of contagion, also analyzed by the authors, the articulation between impurity, purification, and interdiction of contact, as a result of the belief in contagion, embodies the symbolic management of both internal and external dangers. This belief that purification emerges from the interdiction of contact coerces healthy individuals to avoid any physical and social approach with patients and other individuals perceived as dangerous in terms of disease transmission [4].

In this context, it was deemed pertinent to suggest to *Societies* the Special Issue COVID-19 and Social Sciences, justified by the fact that the potential contribution of social sciences is not being sufficiently mobilized. This special issue aims to:

> [. . .] contribute to advancing our understanding of the heuristic capacity of Social Sciences as a fundamental tool in the analysis of cognitive assessments and collective behaviors developed in the pandemic caused by COVID-19 and the implications of the exponential increase of social interactions and spatial, economic, and societal dynamics, at various scales, in the post-pandemic future. [5]

For this Special Issue, COVID-19 and Social Sciences, 14 manuscripts were received, having been selected for publication after several improvements resulting from a rigorous reviewing process, and six papers that focus on different perspectives and are relevant contributions. A brief presentation of each published article follows.

Rudnick [6], in "Social, Psychological, and Philosophical Reflections on Pandemics and Beyond", demonstrates the relevance of several social, psychological, and philosophical issues underlying the COVID-19 pandemic, such as the moral distress of healthcare providers when they have to make decisions about the life or death of patients or related to mental health amplified by quarantine and prophylactic isolation, as well as the need to provide society with additional protection for socially disadvantaged people.

Arias-Maldonado [7], in "COVID-19 as a Global Risk: Confronting the Ambivalences of a Socionatural Threat", proposes to categorize the COVID-19 pandemic as a particular kind of risk that combines premodern and modern features; it takes place in the Anthropocene but is not of the Anthropocene.

Ferreira, Sá, Martins, and Serpa [8] present "The COVID-19 Contagion—Pandemic Dyad: A View from Social Sciences". This manuscript offers a presentation of potential contributions from several specific social sciences, analyzing the analytical potential of social sciences in an informed understanding of the challenges that the COVID-19 pandemic poses to society at the economic, social, and health levels, but which, however, has not been sufficiently mobilized by policymakers.

In turn, Tobin, Mavrommati, and Urban-Rich [9], in "Responding to Social Distancing in Conducting Stakeholder Workshops in COVID-19 Era", offer a contribution on how academic research in the social sciences can, needs to, and must adapt to the compliance with the COVID-19 pandemic control rules. To this end, the authors reflect on real situations of workshops through the mobilization of technology that allows, simultaneously, to satisfy requirements such as social/physical distance. This virtual stakeholder engagement poses potential challenges, several of which are discussed in this article.

Grabowski, Meldgaard, and Hulvej Rod [10], in "Altered Self-Observations, Unclear Risk Perceptions and Changes in Relational Everyday Life: A Qualitative Study of Psychosocial Life with Diabetes during the COVID-19 Lockdown", put forth a study on the psychosocial effects of the conditions for living with a chronic disease such as diabetes in the context of a COVID-19 lockdown in the Danish context.

Finally, Ferri, Grifoni, and Guzzo [11], in "Online Learning and Emergency Remote Teaching: Opportunities and Challenges in Emergency Situations", contribute with an article that aims to analyze the opportunities and challenges of remote education in the context of the COVID-19 emergency, culminating in the analysis of various technological, pedagogical, and social challenges that emerge from this study.

Scientific work does not happen in an isolated or individual way, and hence, we would like to thank all the stakeholders who contributed to the accomplishment of this project that is now offered to the reader. We begin by thanking Dr. Gregor Wolbring, *Societies* Editor-in-Chief, for the confidence placed on us by accepting our call for proposals for this Special Issue. An acknowledgment is due to all authors who submitted manuscripts, reviewers, whose evaluation work was essential and critical to the quality of publications, and the entire editorial office; their professionalism enabled the materialization of this final result.

In conclusion, if any writing becomes the property of the reader after its publication, we would like to believe that this Special Issue COVID-19 and Social Sciences could be yet another vehicle demonstrating the potential of social sciences in developing the understanding how COVID-19 is perceived and experienced. This knowledge is pivotal, for example, in the definition and application of measures to be taken to control the pandemic. Even now that the availability, distribution, and application of vaccines are around the corner, social sciences can make an important contribution to the success of this process.

Funding: This research was funded by the University of the Azores, Interdisciplinary Centre of Social Sciences—CICS.UAc/CICS.NOVA.UAc, UID/SOC/04647/2020, with the financial support of FCT/MEC through national funds and, when applicable, co-financed by FEDER under the PT2020 Partnership Agreement.

Conflicts of Interest: The authors declare no conflict of interest.

References

1. Sá, M.J.; Serpa, S. The global crisis brought about by SARS-CoV-2 and its impacts on education: An overview of the Portuguese panorama. *Sci. Insights Educ. Front.* **2020**, *5*, 525–530. [CrossRef]
2. Sá, M.J.; Serpa, S. The COVID-19 pandemic as an opportunity to foster the sustainable development of teaching in higher education. *Sustainability* **2020**, *12*, 8525. [CrossRef]
3. Aristovnik, A.; Ravšelj, D.; Umek, L. A bibliometric analysis of COVID-19 across science and social science research landscape. *Sustainability* **2020**, *12*, 9132. [CrossRef]
4. Ferreira, C.M.; Serpa, S. Contagions: Domains, Challenges and Health Devices. *Acad. J. Interdiscip. Stud.* **2020**, *9*, 1–14. [CrossRef]
5. Ferreira, C.M.; Serpa, S. Special Issue "COVID-19 and Social Sciences" 2020. Available online: https://www.mdpi.com/journal/societies/special_issues/COVID-19_social_sciences (accessed on 11 December 2020).
6. Rudnick, A. Social, psychological, and philosophical reflections on pandemics and beyond. *Societies* **2020**, *10*, 42. [CrossRef]
7. Arias-Maldonado, M. COVID-19 as a global risk: Confronting the ambivalences of a socionatural threat. *Societies* **2020**, *10*, 92. [CrossRef]
8. Ferreira, C.M.; Sá, M.J.; Martins, J.G.; Serpa, S. The COVID-19 contagion–pandemic dyad: A view from social sciences. *Societies* **2020**, *10*, 77. [CrossRef]
9. Tobin, C.; Mavrommati, G.; Urban-Rich, J. Responding to Social Distancing in Conducting Stakeholder Workshops in COVID-19 Era. *Societies* **2020**, *10*, 98. [CrossRef]
10. Grabowski, D.; Meldgaard, J.; Hulvej Rod, M. Altered self-observations, unclear risk perceptions and changes in relational everyday life: A qualitative study of psychosocial life with diabetes during the COVID-19 lockdown. *Societies* **2020**, *10*, 63. [CrossRef]
11. Ferri, F.; Grifoni, P.; Guzzo, T. Online learning and emergency remote teaching: Opportunities and challenges in emergency situations. *Societies* **2020**, *10*, 86. [CrossRef]

Publisher's Note: MDPI stays neutral with regard to jurisdictional claims in published maps and institutional affiliations.

© 2020 by the authors. Licensee MDPI, Basel, Switzerland. This article is an open access article distributed under the terms and conditions of the Creative Commons Attribution (CC BY) license (http://creativecommons.org/licenses/by/4.0/).

Concept Paper

Social, Psychological, and Philosophical Reflections on Pandemics and Beyond

Abraham Rudnick

Department of Psychiatry and School of Occupational Therapy, Dalhousie University, Halifax, NS B3H 4R2, Canada; harudnick@hotmail.com

Received: 26 April 2020; Accepted: 30 May 2020; Published: 1 June 2020

Abstract: This conceptual paper presents social, psychological and philosophical (ethical and epistemological) reflections regarding the current (COVID-19) pandemic and beyond, using an analytic and comparative approach. For example, Taiwan and Canada are compared, addressing Taiwan's learning from SARS. Suggestions are made in relation to current and future relevant practice, policy, research and education. For example, highly exposed individuals and particularly vulnerable populations, such as health care providers and socially disadvantaged (homeless and other) people, respectively, are addressed as requiring special attention. In conclusion, more reflection on and study of social and psychological challenges as well as underlying philosophical issues related to the current pandemic and more generally to global crises is needed.

Keywords: education; pandemic; philosophy; policy; practice; psychology; research; social

1. Introduction

Societies are measured in part in relation to how they rise to the occasion of collective crises and learn from them. For example, both Taiwan and Canada (specifically Toronto) were similarly directly impacted by the Severe Acute Respiratory Syndrome (SARS) pandemic and related nosocomial (hospital-based) viral transmission a couple of decades ago [1], yet it seems that Taiwan learned from that to prepare well for such pandemics, whereas Canada (including Toronto) did not [2]. The current Coronavirus Disease 2019 (COVID-19) pandemic is such a crisis and raises various problems that are insufficiently addressed to date (such as the impact of international travel on global health), some of them reflective of underlying social and other challenges across the world [3]. In addition to medical and technological problems, social as well as psychological problems and underlying philosophical (particularly ethical and epistemological) challenges have to be better addressed to further improve the approach to this pandemic and arguably to future pandemics and other global crises. In this conceptual paper, I use an analytic [4] and comparative [5] approach to present related social, psychological, and philosophical issues, using my experience and expertise as a social scientist and health researcher [6], a clinically practicing psychiatrist, a health care administrator [7,8], and a philosopher of health and related care [9,10]. I conclude with practice and policy as well as research and education suggestions.

2. Social and Psychological Reflections

The current pandemic poses important social challenges. For example, many people have been laid off work temporarily or permanently during the pandemic due to an insufficient workload, such as in the service sector. Unemployment is associated with disrupted mental well-being [11] and with other personal as well as societal disruptions such as poverty, crime, and more. The most vulnerable to such disruptions are typically people who are already disadvantaged, such as those from lower socioeconomic strata and many retired people. Hence, the general population, and particularly vulnerable populations such as socially disadvantaged people (homeless individuals and others), may require particular attention during and soon after the pandemic to try to ensure that they are at least not further disadvantaged. Another example is the expected political disruption during a pandemic, particularly in countries where the regime is not democratically robust (such as in Israel where the prime minister is allowed to stay in office in spite of incurring criminal charges [12]). In such countries, some people may use the opportunity during the pandemic to disorganize society or to further restrict the general public and/or special social groupings that are considered by them as socially undesirable (such as racialized minorities and others). Such disorganization and restriction can further disrupt personal and societal well-being during the pandemic (and after it if the disruptive political changes remain in place). Hence, the general public and/or special social groupings that are considered socially undesirable by some may require particular attention during and after the pandemic to try to support them in relation to pandemic-related disruptive political change.

The current pandemic also poses psychological challenges. For example, (self) quarantine and isolation may seem similar; but (self) quarantine is separation for people who were actually or plausibly exposed to a contagious disease (such as from international travel) but are not confirmed to be infected, whereas isolation is for people who are infected with a contagious disease [13]. As such, (self) quarantine may seem less stressful, not only because the person is presumably not infected, but also because the person is supposedly in control of their quarantine. Yet the stress of not being tested (as in many jurisdictions only symptomatic people or people who have been in contact with infected people are tested) may worsen the (self) quarantined person's stress. Also, the social pressure—and the legal requirement in an increasing number of jurisdictions—to (self) quarantine may reduce the person's sense of control and even generate distress related to the discrepancy between social expectations and individual entitlement to freedom of movement (in jurisdictions where that is legally supported). Hence, the highly prevalent psychological needs for certainty and for sense of control are not easily addressed in self-quarantine and may require particular attention during the pandemic to facilitate mental well-being of (self) quarantined people. Another example is the likely loss of trust in people who are physically close (and personally significant) to a person in case they are either infectious (while asymptomatic) or are not careful enough in trying to prevent being infected. Such a pervasive loss of trust may deeply disrupt people's mental well-being and functioning, particularly if they are already vulnerable such as having an insecure attachment style [14]. Hence, the universal psychological need for trust is not easily addressed with family and friends during the pandemic, particularly in relation to emotionally vulnerable people, and may require particular attention during the pandemic to facilitate their mental well-being and functioning. These and other psychological challenges related to the pandemic period may last beyond it, especially if there were personally traumatic events during it, such as forced self-quarantine by authorities and betrayal of trust by (personally) significant others. These challenges may require special attention after the pandemic to facilitate mental well-being and functioning of people who are identified as having—or being at high risk of having—pandemic-related mental problems after the pandemic.

3. Philosophical (Ethical and Epistemological) Reflections

Some of the pandemic-related social and psychological issues are associated with underlying ethical issues. An example is the scarcity of health care resources, which is rampant during the current pandemic, as it has been during some other pandemics, such as the Spanish flu pandemic (when human health resources—particularly physicians and nurses—in the United States were depleted due to their deployment abroad near the end of World War I [15]); decisions about which treatable patients to exclude from treatment—such as ventilation—can cause moral distress and other disruption to health care providers. Hence health care providers may require particular attention during and after the pandemic to address their moral distress. Another example is the common—personal and social—expectation during the pandemic that individuals help others, above and beyond what is expected in more ordinary times. Although ordinary ethics would consider that as supererogatory, i.e., laudable but not required morally, during extraordinary—such as pandemic—times, extraordinary moral conduct may be expected if not required, involving increased individual moral responsibility [16]—including for others' plight even if there is no preset relationship between them and the individual expected to help them. Hence, the general public may require particular attention during and after the pandemic for emotional and practical support in relation to such extraordinary moral conduct expectations.

Some of the pandemic-related social and psychological issues are also associated with underlying epistemological issues. For example, individual and collective behavior impact biological aspects of the pandemic such as rate of transmission, yet robust evidence on that is difficult to obtain due to the lack of availability of randomized controlled trials in such circumstances. Other approaches to generate robust evidence are needed in these circumstances, such as studies comparing naturally variant sites and populations and sufficiently matched samples, recognizing that comparison is key to any inquiry [5]. Hence, researchers may require particular attention during the current pandemic and in preparation for future pandemics and other global health crises to optimize their research methodology for such circumstances. Another example is the common misunderstanding by lay people of what is robust evidence, which may lead to their unsafe behavior or alternatively to their overly cautious behavior during the pandemic. This may pose unnecessary personal harm and public risk (due to increased transmission of infection) or alternatively unnecessary personal restriction and social disruption (due to unnecessary reduction of work and other activities), respectively. Hence, the general public may require particular attention during the current pandemic (and arguably at all other times) to enhance lay people's critical thinking and knowledge about evidence and other relevant aspects of rigorous inquiry such as health research.

4. Conclusions

Social, psychological, and underlying philosophical issues that are pandemic-related may have a considerable and lasting impact on societies and on particular individuals. Some related practice suggestions are to address the moral distress of health care providers who have to make particularly difficult—sometimes life or death—decisions due to very scarce health care resources, and to provide additional emotional support such as to (self) quarantined people and to people who have pre-pandemic mental challenges (preferably provided by their significant others and/or mental health care providers). Some related policy suggestions are to secure additional income support for socially disadvantaged people during and soon after the pandemic, and to provide additional protections for special social groupings that are considered socially undesirable by some if the pandemic results in disruptive political change (that may last after the pandemic). Some related research suggestions are to study societal preparation for pandemics, perhaps learning from positive deviance such as Taiwan's successful preparation for the current (COVID-19) pandemic based on its experience with the SARS pandemic nearly 20 years ago [1]. Some related education suggestions are to train the general public as well as health care providers and other first responders in advance in responsible behaviors that protect them and others during a pandemic and other challenging times. More reflection on and study of social and psychological challenges as well as underlying philosophical issues related to the current pandemic, and more generally to global crises, is needed.

Funding: This research received no external funding.

Conflicts of Interest: The author declares no conflict of interest.

References

1. McDonald, L.C.; Simor, L.E.; Su, I.-J.; Maloney, S.; Ofner, M.; Chen, K.-T.; Lando, J.F.; McGeer, A.; Lee, M.-L.; Jernigan, D.B. SARS in Healthcare Facilities, Toronto and Taiwan. *Emerg. Infect. Dis.* **2004**, *10*, 777–781. [CrossRef] [PubMed]
2. Schuchman, M. We Can Learn from Taiwan on How to Fight Coronavirus. *The Star*. Available online: https://www.thestar.com/opinion/contributors/2020/03/16/we-can-learn-from-taiwan-on-how-to-fight-coronavirus.html (accessed on 16 March 2020).

3. Zizek, S. Pandemic! COVID-19 Shakes the World. Available online: https://www.orbooks.com/catalog/pandemic/ (accessed on 16 March 2020).
4. Yehezkel, G. A Model of Conceptual Analysis. *Metaphilosophy* **2005**, *36*, 668–687. [CrossRef]
5. Rudnick, A. A Philosophical Analysis of the General Methodology of Qualitative Research: A Critical Rationalist Perspective. *Health Care Anal.* **2014**, *22*, 245–254. [CrossRef]
6. Rudnick, A.; Forchuk, C. *Social Science Methods in Health Research*; Sage: New Delhi, India, 2017.
7. Rudnick, A. Principled Physician (and Other Health Care) Leadership: Introducing a Value-Based Approach. *Can. J. Physician Leadersh.* **2014**, *1*, 7–10.
8. Pallaveshi, L.; Rudnick, A. Development of Physician Leadership: A Scoping Review. *Can. J. Physician Leadersh.* **2016**, *3*, 57–63.
9. Rudnick, A. *Bioethics in the 21st Century*; InTech: Rijeka, Croatia, 2011.
10. Rudnick, A. *Recovery from Mental Illness: Philosophical and Related Perspectives*; Oxford University Press: Oxford, UK, 2012.
11. Farré, L.; Fasani, F.; Mueller, H. Feeling Useless: The Effect of Unemployment on Mental Health in the Great Recession. *IZA J. Labor Econ.* **2018**, *7*, 1.
12. Silverstein, R. Israeli Court's Nod to Netanyahu-Gantz Deal Cheapens Democracy. *Middle East Eye*. Available online: https://www.middleeasteye.net/opinion/approval-israels-top-court-netanyahu-gantz-deal-cheapens-democracy (accessed on 9 May 2020).
13. Centers for Disease Control and Prevention. Quarantine and Isolation. Centers for Disease Control and Prevention, National Center for Emerging and Zoonotic Infectious Diseases, Division of Global Migration and Quarantine, 29 September 2017. Available online: https://www.cdc.gov/quarantine/index.html (accessed on 9 May 2020).
14. Rodriguez, L.M.; DiBello, A.M.; Øverup, C.S.; Neighbors, C. The Price of Distrust: Trust, Anxious Attachment, Jealousy, and Partner Abuse. *Partn. Abus.* **2015**, *6*, 298–319. [CrossRef] [PubMed]
15. Ott, M.; Shaw, S.F.; Danila, R.N.; Lynfield, R. Lessons Learned from the 1918–1919 Influenza Pandemic in Minneapolis and St. Paul, Minnesota. *Public Health Rep.* **2007**, *122*, 803–810. [CrossRef] [PubMed]
16. Rudnick, A. Moral Responsibility Reconsidered: Integrating Chance, Choice and Constraint. *Int. J. Philos.* **2019**, *7*, 48–54. [CrossRef]

© 2020 by the author. Licensee MDPI, Basel, Switzerland. This article is an open access article distributed under the terms and conditions of the Creative Commons Attribution (CC BY) license (http://creativecommons.org/licenses/by/4.0/).

Article

COVID-19 as a Global Risk: Confronting the Ambivalences of a Socionatural Threat

Manuel Arias-Maldonado

Department of Political Science, University of Málaga, 29016 Málaga, Spain; marias@uma.es

Received: 25 October 2020; Accepted: 25 November 2020; Published: 26 November 2020

Abstract: On the face of it, the COVID-19 pandemic seems to fit into the risk society framework as a danger that is produced by the modernization process in its global stage. However, coronaviruses are a very particular kind of risk which risk theory does not properly explain. In fact, there is no single perspective on risk that offers a fully satisfactory account of the SARS-CoV-2, despite all of them having something valuable to contribute to the task. This paper attempts to categorize the COVID-19 pandemic as a particular kind of risk that is not adequately explained with reference to the risk society or the new epoch of the Anthropocene. On the contrary, it combines premodern and modern features: it takes place in the Anthropocene but is not of the Anthropocene, while its effects are a manifestation of the long globalization process that begins in antiquity with the early representations of the planet as a sphere. If the particular identity of the disease is considered, COVID-19 emerges as the first truly global illness and thus points to a new understanding of the vulnerability of the human species qua species.

Keywords: risk; COVID-19; Anthropocene; modernization; globalization; disease identity

1. Introduction

What kind of risk is the COVID-19 pandemic? How can the social sciences help us to understand the rapid spread of the coronavirus SARS-CoV-2 and its impact across the globe? Which are the available theoretical frameworks for doing so and how well do they apply to zoonotic infections turned global? Is perhaps the Anthropocene, as a brand-new hypothesis about the human impact on the natural world and its unintended effects, the most appropriate explanation?

These are the questions that this paper will deal with. I depart from the assumption that despite the spectacular quality of the COVID-19 pandemic, this was neither an unprecedented nor an unforeseeable risk. Natural scientists have been warning for some time now that the presence of a large reservoir of SARS-CoV-like viruses in horseshoe bats, together with the culture of eating exotic mammals in southern China, is a time bomb [1]. The same fear was expressed in a piece published in *Nature* five years ago, in which the possibility of a human emergence was explicitly formulated as a way to call the attention of public authorities [2]. Even among economists there was someone who alerted, just a month before the outbreak in Wuhan came to be publicly known, about the potentially high cost of a global pandemic to prevent which not much was being done [3].

They were all quite right, although the advantages of retrospection must also be considered. However, pandemics are hardly an unknown historical phenomenon. Zoonotic spillovers, in which an infectious virus jumps from non-human animals to humans, are not new either. It should be remembered that so-called Spanish Flu killed between 50 and 100 million people around the world in 1918–1919, becoming the most globalized epidemic in history [4], and despite having been mostly forgotten, the Asian Flu in 1957 killed around 2 million people all over the world and reduced US economic growth by 10% in the first quarter of the following year [5], whereas the Hong-Kong Flu caused the death of a million people in 1969, including 25,000 French poeple just in December [6]. Later, zoonotic viruses come one after another: the Lassa (1969), the Ebola (1976), the VIH-1 and VIH-2 that cause AIDS (1981, 1986), the Hendra (1994), avian flu (1997), Nipah (1998), West Nile (1993), swine flu (2009), and finally SARS-CoV-2. David Quanmen, a science journalist specialized in the phenomenon, wrote a couple of years ago that the word zoonosis was going to be around for some time now [7]. Just when some historians were labeling the last one hundred years as the pandemic century [8], COVID-19 seems to herald the continuation of an infectious epoch.

Yet there is no single mention of pandemics in the late Ulrich Beck's account of the global risk society, which was published eleven years ago [9]. This omission is shocking, although nobody had noticed it until COVID-19 began its unstoppable global spread. What does this say about risk society theory? Is it just a distraction on the part of Beck, or perhaps viruses do not have a place in his account of the relation between modernity and risk? If that is the case, which other approaches to risk may assist us in understanding pandemics as particular social threats? Such diagnoses matter, because the way in which societies relate to their environments has much to do with their own self-understanding. This resonates with Luhmann's assertion that there exists a strong critical potential in the analysis of the way in which a society confronts misfortune, as the latter makes it possible to see more thoroughly the reverse of their normality [10]. On such occasions, societies resemble the sick body in which a fluorescent liquid is introduced when a tomography is performed—their inner functioning can be briefly observed in more detail.

Hence this paper deals with the ambiguous relation between COVID-19 and modernity. It does so through the lenses provided by the category of risk—including ecological risks and the new theoretical and symbolic framework provided by the Anthropocene. Against the idea that the pandemic is a typically modern event caused by a predatory view of socionatural relations in the context of a capitalistic-driven globalized world, I suggest that the former is a rather primitive threat that has accompanied human populations ever since they have existed as such. Needless to say, pandemics reflect the features of the age in which they take place. In the case of COVID-19, globalization and social acceleration serve to explain its rapid spread as much as the unprecedented swiftness with which the virus has been decoded and a number of vaccines have been announced. In fact, the pandemic is less the result of a *failed* modernity than it is the outcome of a *lack* of modernity. On the one hand, the virus overcomes the species barrier in a country, China, where food security is notoriously lacking and the flow of information is restricted for political reasons. On the other, most Western countries have resorted to rather primitive strategies of contention based on lockdowns and have proven themselves unable to display a data-driven, sophisticated approach to the pandemic.

The article is organized as follows. Section 3 offers an overview of the relation between globalization, risk, and pandemics. Section 4 discusses risk society and the Anthropocene as theoretical frameworks that may help to explain the COVID-19 pandemic. Section 5 employs cultural approaches to risk in order to highlight the difference between perceived and objective risks. Section 6 deals with the metaphorical dimension of COVID-19, suggesting that this may be the first disease that is perceived as affecting human beings as a totality, irrespective of their ethnic or social belonging. Finally, Section 7 suggests the need to rethink global risks after the corona pandemic, so that a more balanced account of the relation between potential threats and materialized disasters is achieved. Section 8 serves as a conclusion.

2. Methodology

This article uses a qualitative methodology, as it seeks to understand the nature and meanings of the COVID-19 pandemic with the tools of the social sciences (sociological risk theory, political theory) as well as with those of other, typically interdisciplinary, academic endeavors (environmental studies, epidemiology and other natural sciences). The article does not involve an empirical effort, although it benefits from the work of empirical social and natural scientists.

3. Globalization, Risk, and Pandemics

The COVID-19 pandemic has impacted on a globalized world in which the circulation of people, goods, and information had reached an unprecedented magnitude and speed—despite the political backlash that started to take shape in the aftermath of the Great Recession. Yet the history of globalization is interlocked with that of epidemics, as the effects of the Spaniards' landing on the health of native Americans back in the 16th century came to show; the sudden contact between two human populations that had been separated for millennia caused a number of lethal epidemics that facilitated the European conquest of the Americas [11]. Thus, the seminal fact of modernity's globalizing praxis is attached to a warning about the epidemic potential of the world's shrinking. Something of the sort had happened in continental Europe during the Black Death in the 14th century, albeit at a lesser scale. On its part, the novel coronavirus SARS-CoV-2 has not provoked an ethnic catastrophe, nor has it thrived in the trenches of a devastating war—it has travelled from China to Europe as one of the trade goods in post-communist globalization.

The link between globalization and pandemics is not accidental. The latter are called crowd diseases for a reason: the epidemic transmission of a virus requires an abundant population and that explains why many viral diseases are relatively recent. Although traces of smallpox have been found in Egyptian mummies, polio is not documented until the 19th century and AIDS appeared at the end of last century [12]. It is not surprising, then, that epidemics do not exist before big human populations, nor that the viruses take advantage of the commercial routes through which individual carriers circulate together with their merchandise [13]. Yet the homogenizing of the global biota that begins with the so-called Columbian exchange after the Spanish colonization of America produces in turn an ambiguous effect on the infectious vulnerability of human populations. Environmental historian Alfred Crosby [11] has highlighted its negative impact on genetic diversity:

> The flora and fauna of the Old and especially of the New World have been reduced and specialized by man. Specialization almost always narrows the possibilities for future changes: for the sake of present convenience, we loot the future. (...) The Columbian exchange has left us with not a richer but a more impoverished genetic pool.

Biodiversity loss has also been singled out by biologist Edward O. Wilson [14] as a threat for the future of the human species and, consequently, one of the major threats of the Anthropocene. The Anthropocene is the name given to the epoch in which natural systems at a planetary level have been disrupted by human activity. Anthropogenic climate change is perhaps the best-known manifestation of this phenomenon, together with ocean acidification or biodiversity loss. A group of geologists claim that this should actually be recognized as a new *geological* epoch, insofar as the change induced by human activity would have left a trace in the fossil record [15]. Regardless of the decision eventually adopted by the International Commission on Stratigraphy, the Anthropocene can also be taken as a historical epoch, a rupture in socionatural relations that is sustained in biological and ecological evidence [16].

Nevertheless, it is not clear whether ecological homogenization worsens the lethality of epidemics. On the one hand, humanity as a whole becomes more vulnerable to an extremely aggressive super virus insofar as the lack of human genetic diversity would facilitate the sinister task of the germ. The more lethal a virus is, however, the less easily does it spread—carriers who die often cannot communicate the disease so easily as those who can live with it for a longer period. On the other hand, it becomes more unlikely that geographically distant populations can inflict on each other the kind of acute harm that Spanish *conquistadores* unintentionally inflicted on native Americans by transmitting to them diseases against which the latter had no immune defense. A different matter is that a densely interconnected world can experience a great disruption by the quick global spread of a highly contagious and moderately lethal virus—as the SARS-CoV-2 that causes COVID-19 has proven to be [17].

The global nature of the virus is expressed in that it has a local origin that quickly becomes transcended by an accelerated diffusion, so that a situated phenomenon is rapidly turned into a planetary event: zoonosis infections begin *somewhere* but may end up happening *everywhere*. The pandemic presents the appearance of that which Virilio [18] designates as an integral accident, which is the one capable of integrating a whole heap of incidents and disasters through chain reactions. This alarmist view of modernity places emphasis on the fatal consequences of globalization, described as a sort of "voyage to the center of the Earth" in the gloomy obscurity of a temporal compression that definitively looks down upon the habitat of the human race [18]. The epidemic circle is thus completed: the Columbine tragedy repeats itself as a zoonotic farce that empties out the malls in a world whose confines have already been thoroughly explored. Most recently, Andreas Malm has explained the pandemic as a consequence of Western imperialism: a capitalistic oil-fueled economy that thrives on natural destruction intrudes in the wild and forces pathogens to leap towards us [19].

Despite the support that it seems to receive from the pandemic, the apocalyptic view of globalization denotes a narrow understanding of this complex human phenomenon. According to Sloterdijk, globalization is not a modern event, but rather is a far more logically and historically powerful process that begins when the Greeks started to conceive the world as a totality [20]. A similar point can be found in Carl Schmitt's idea of the spatial revolution, which designates moments in which new lands and seas are incorporated into human collective consciousness, thus transforming the locations in which its historical existence takes place [21,22]. Throughout this process, global risks emerge—modernity is the time in which thinking about the globe is supplemented by action. The expansion of capitalism is key for understanding this change, but it is not the only factor at play: modernity involves a gradual liberation from previous constrictions, and this allows the exploration of a world that can now be economically exploited, religiously evangelized or culturally civilized [20]. It is in this context that insurance is born, as an institution that protects entrepreneurs against the dangers of seafaring—in the event of an accident, those who pay their fees are proportionately compensated by the insurance company. Business investments are thus made safer and the notion of risk acquires its modern meaning.

Despite the obscurity that surrounds its origin, the word risk may come from the Arabic *risq* (meaning wealth or good luck), from the Greek *rhiza* (cliff) or the Latin *resegare* (which means cutting with a single stroke). It might have been firstly used to designate the danger of navigating close to the rocks [23]. Arguably, that is what humanity has always done: navigating close to the rocks. Yet at a given point in history individuals and groups started to calculate the likeliness of accidents and other contingencies, thus showing confidence in their ability to handle their relations with the environment. Risk calculation produced a number of legal institutions, administrative customs, economic interests and technological appliances, all of which gave shape to modern capitalistic societies—from the assurance provided to the investor to the sickness leave enjoyed by the worker. The scientization of risk began in the 18th century, in connection with the development of probability. In the following century, positivism reinforced the trend so much that some people believed that chance does not play any role in shaping social reality; objective knowledge should lead to the taming of chance [24]. From the vantage point of the present, it may be surprising that a century in which three cholera epidemics took place in Europe—Hegel himself died because of the third one—exhibited such an unrestrained faith in reason. Since then, risk becomes less a property of the world than a property of the human knowledge of the world: there is risk where there is not yet enough knowledge [25].

There are reasons to believe that this worldview, in which a radically indeterminate world becomes calculable through the exercise of reason and conquerable through insurance, still permeates late modern societies. It remains to be seen whether the COVID-19 pandemic will change this, but it does not seem likely—confidence in the ability of the scientific system to find a vaccine against the disease has not been shaken throughout the pandemic. Such confidence resembles a faith, a typically modern belief that is however grounded on past achievements and not just on false promises.

4. From Risk Society to the Anthropocene: The Search for Explaining COVID-19

The speed at which the coronavirus SARS-CoV-2 has spread around the globe has been unanimously interpreted as a consequence of the dense interconnection that binds national societies to each other. This is just another manifestation of the dark side of modernity, which usually expresses itself in the form of collateral damages. A most spectacular instance of this structural feature is anthropogenic climate change, but many others can be added to this sinister lot—nuclear hazards, food crisis, technological collapses. In the words of Virilio, the new century inaugurates the paradox of the failure of success, for it is the success of progress that provokes disaster [18]. The assumption that modernity creates its own threats is naturally at the center of Ulrich Beck's risk theory, which was formulated by the late German sociologist in the mid-1980s and has now been signaled as an obvious starting point for making sense of the coronavirus pandemic [26]. This feeling is summarized in the statement that Ulrich Beck's 'risk society' appears to be taking on new forms in current times [27]. The pandemic is thus seen as a risk that has materialized, a threat that has inflicted a great harm on current societies across the planet, thus validating, empirically, in spectacular fashion, Beck's theory about the reflexive quality of modernization.

In the company of fellow sociologists such as Anthony Giddens and Scott Lash, Beck deals with risk from a macrosocial perspective, putting the latter in connection with key features of modern societies: individualization, detraditionalization, reflexivity [26,28–30]. Rather than dissolving themselves in postmodernity, Western societies entered into a new stage of modernity in which some of the latter's features are radicalized. These societies are reflexive precisely because they are confronted with the unintended consequences of their development. According to Beck [9],

> Risk represents the perceptual and cognitive schema in accordance with which a society mobilizes itself when it is confronted with the openness, uncertainties and obstructions of a self-created future and it is no longer defined by religion, tradition or the superior power of nature but has even lost its faith in the redemptive power of utopias.

In this regard, our epoch is defined by new kinds of potential dangers hitherto unknown, namely those that do not obey the vagaries of fortune and instead originate in the modernization process. Their source is social, not natural—the problem lies in DDT rather than in the locust plague. Moreover, some of them are so huge that they threaten human survival. For the same reason, the logic of insurance does not apply: there is no company that can compensate a series of nuclear accidents. Foreseeing risks thus becomes a public task, as the state is in charge of counteracting an industrial fatalism that threatens to spoil the fruits of modernization [29].

It should be noted that a *risk* is not the same as a *catastrophe*—risks are potential catastrophes and thus their anticipation. Strictly speaking, we are referring to threats or dangers of a particular kind, namely those that are socially originated and depend on human decisions. That is why Beck speaks of manufactured uncertainties which are socially created, collectively imposed and individually inescapable [31]. In his view, modern societies are constantly shaken by the global anticipation of global risks: financial crises, climate change, terrorist attacks. As risk is a possible disaster that has not taken place yet, imagination is essential—societies formulate self-refuted prophecies that require the adoption of precautionary measures [9]. Otherwise, the risk can give way to catastrophe. For Giddens, modernity is double-faced and spontaneously produces a risk culture [32]. How likely is it that what should not happen finally happens? Potential threats are an expression of contingency and uncertainty, and it is through the concept and culture of risk that late modern societies express the impossibility of extending the dominion of reason to each and every sphere.

What about the COVID-19 pandemic? Does it fit into the category of global risk, as Beck elaborates it? Is the coronavirus a product of risk society? Although there is the temptation of taking this for granted, the answer is not straightforward. The pandemic certainly participates in some of the features that Beck attributes to modern risks—it has a local origin and global consequences, its potential effects are not calculable and barely indemnified. Viruses are also threats that can be imagined without necessarily happening, as the number of existing viruses does not correspond to the number of epidemic outbreaks.

However, a virus is hardly a *consequence* of modernity. On the contrary, it should be catalogued as a *remainder* of premodernity—a primitive threat that is related to the animal side of human beings and thus to a biological condition that cannot be suppressed. Likewise, zoonotic viruses have not disappeared in the Anthropocene and it is even likely that they become more frequent in the current stage of socionatural relations [33]. Yet these entities are old inhabitants of the planet and they already represented a danger for human beings in the Holocene. They were in fact a greater danger, as modern virology did not exist yet and there was no way to fight them. That is why the statement that COVID-19 is the disease of the Anthropocene does seem a bit hasty [34]. A different matter is that viruses seemed to have been forgotten in the face of other global risks. As the 20th century demonstrates, though, they really never disappeared. Granted, the rapid spread of the virus cannot be explained without the globalization process that increases mobility and connects parts of the planet that used to be isolated from each other. This pandemic is thus a combination of the old and the new, the product of a zoonotic spillover that reproduces a well-known human vulnerability in a contemporary setting.

On the other hand, the coronavirus is not a social creation nor can it be attributed to any particular human decision. Are pandemics a natural risk then? This is another hard question. Virilio distinguishes between *natural* and *artificial* accidents: the former originate in the non-human world and the latter result from the innovation of a motor or of some substantial material [18]. Yet if the ship entails the invention of wrecks and nuclear energy created the risk of neutron leakage, which is the human invention that produces the threat of zoonotic spillover? As a matter of fact, it does not exist. The potentially infectious contact between humans and animals is a natural circumstance, and as such there is nothing remotely modern about it. If the anthropos of the Anthropocene is to be understood as a technological subject who inhabits a multifaceted technosphere that includes the industrial production of meat and food additives that transform human bodies or those of cultivated livestock [35], then all that is known so far about the source of SARS-CoV-2 speaks of an anthropos that maintains a rather unmediated relation with animals. Instead of food additives, wild animals as captured and sold in wet markets may contain zoonotic viruses. Dyer-Whiteford rightly speaks of a "planet factory" to describe a world in which a world-market crisscrossed by supply chains operates by producing and destroying the biosphere itself [36]. Had the virus been originated in a meat processing plant, the causal and symbolic connection to the Anthropocene would be much clearer.

Admittedly, this is not to deny that the modern organization of socionatural relations may facilitate contagion and accelerate the global communication of the disease. Society and nature are now more firmly interwoven than in the past:

> Nature has become an integral part of societal reproduction both in its positive guise as a provider of the material assets of social life and in its negative dimensions as a risk for our health, safety and the possibilities for future development [37].

In this particular sense, the pandemic can actually be categorized as a *socionatural* rather than just a *natural* risk. As Ferreira et al. point out, a pandemic is always a point of articulation between natural and social determinations [38]. Corona resembles hurricanes: although they have always existed, their greater current intensity can be related to anthropogenic climate change [39]. Likewise, changing patterns of animal migration caused by global warming or habitat invasion may significantly affect how zoonosis takes place. A pathogen has already been identified as the first mycotic disease produced under the new planetary conditions—the *candida auris* discovered simultaneously in three different continents [40]. Whether we are witnessing a microbial insurgency, only time will tell, and yet the COVID-19 pandemic does not seem reason enough to speak on such terms [41,42]. Humanity's recorded history, after all, is full of epidemics originated in some non-human animal. Further, given that the sedentary organization of human beings is a precondition for the massive spread of viruses, it makes sense to label epidemics as *dangers of the Holocene* rather than as *risks of the Anthropocene*—provided that risks as such are confined to modernity.

Naturally, this much depends on the moment in which the beginning of the Anthropocene is located. From the viewpoint of geologists, the beginning of the Anthropocene cannot be dated before industrialism has produced its effects—otherwise there are no global, synchronous effects to observe in the fossil record [15]. As much as the agricultural revolution associated to human sedentariness gives way to an intensification of socionatural relations in the Holocene, it is not enough to produce a disruption of planetary systems. Sedentariness, though, brought about crowd diseases and thus epidemics. In sum, if Anthropocene risks are those that originate from anthropogenic changes in key functions of the Earth system, whose emergence is due to the evolution of intertwined social-ecological systems often featuring inequality and injustice, and which exhibit complex cross-scale interactions that potentially involve Earth system-tipping elements [43], then the COVID-19 pandemic should not be categorized as an Anthropocene risk.

To be sure, the conditions that created the Anthropocene may increase the risk of zoonotic spillover in the near future, and it is undeniable that the conditions of late modernity accelerate the global spread of the disease. However, it does not follow that neither risk as described by the sociologists of reflexive modernization nor the Anthropocene as it is commonly understood provide the best explanation for pandemics—and for the COVID-19 pandemic in particular. Unsurprisingly, the sharp distinction between natural pre-modern risks and manufactured modern risks has been one of the weak spots of risk society theory from the outset [44]. Many modern risks exhibit natural features, whereas many premodern threats have social components. The claim that risk society begins at the point in which nature disappears is just too categorical. It has to be considered that the human adaptation to the environment, which is an aggressive adaptation in that it involves the transformation of the environment to satisfy human needs, is hardly a novelty. What has changed in the last two centuries is the magnitude of that transformation.

In this regard, the pandemic reflects less an inherent feature of modernity than the contrast between different modernities—or rather, different rhythms of modernization. Industrialization, imperial colonialism and globalization explain that the Eastern epidemics of the 19th century reached Europe, that the Spanish Flu was a global episode, and the fact that in the last two decades several epidemics originating in Asia have spread across the Western world. It is not by chance that the two significant outbreaks of flu that were communicated to Europe and the US during the Cold War—1957 and 1969—originated in Hong-Kong under British rule, and it goes without saying that China's development in the last four decades represents an important portion of the Great Acceleration of the Anthropocene—that which begins after the second world war. Still in the wild stage of modernization, Asian societies are a frequent source of zoonotic viruses, and this reflects a number of cultural practices as much as a more casual approach to risks associated to food security. This is another paradox of development—while fighting poverty increases pollution, it also increases the likeliness of zoonotic infection. Thus the COVID-19 pandemic partially reflects a kind of risk that is produced by the incorporation to modernity into an interconnected world in which communicable threats are easily communicated.

In sum, although Beck's account of risk illuminates some aspects of the COVID-19 pandemic, the latter's explanation cannot entirely rely on this theoretical framework. The same goes for the Anthropocene hypothesis, which has been suggested as a major explanation for the occurrence of the pandemic—and which can be seen in itself as a major consequence of modernity, as the human colonization of nature acquires a new dimension in the industrial era. The problem with such approaches to COVID-19 is that the search for a novel explanation may lead to overlooking that which is not new in a communicable disease originating in a zoonotic spillover. Despite the grand talk about emerging viral diseases, viral outbreaks have always emerged that way, and have always produced previously unknown diseases. The novelty lies in how fast the disease has spread globally—a relative novelty, though, taking the Spanish Flu into consideration. Admittedly, the COVID-19 pandemic cannot be fully explained without resorting to either risk society theory or the circumstances created by the Anthropocene. The coronavirus SARS-CoV-2 is a hybrid: a pre-modern threat that shares some features with modern risk, as well as a danger of the Holocene that is amplified in the Anthropocene. Yet it is not a manufactured risk, nor one that may be explained as a side-effect of the way in which modern humans relate to nature—there is nothing particularly modern about been infected by eating a bat or a pangolin after having captured them in the wild or having purchased them alive or just killed in a wet market.

In the wider field of risk studies, though, there are valuable analytical tools beyond risk society theory—ranging from the scientific–cognitive perspective that seeks to identify observable risks to the culturalist approaches that focus on the social and cultural contexts within which risks are perceived and experienced, to the governmentality theory that sees risk as a way of disciplining individuals in the name of safety. Let us now turn to them.

5. The Pandemic as a Perceived Risk

Is there such thing as a *real* risk, or are they all somehow *imagined* by particular communities? Can risks be objectified, or rather do they result from a process of social construction that determines the individual perceptions of them? In this particular case, have societies and individuals reacted to the spread of coronavirus in a rational way? Or perhaps have social and cultural factors been decisive in shaping such responses in each case?

Despite the skeptical stance initially adopted by some observers, such as philosopher Giorgio Agamben [45] when he likened the pandemic to a flu and warned against the threat of biopolitical authoritarianism, there are few doubts left about the reality of the pandemic. This has nothing to do with the idea that risks are the anticipation of disaster, an exercise in pre-emptive imagination on the part of modern societies. Yet risks are not *imaginary*, but rather *imagined*—even though some can certainly belong to the former category and refer to implausible threats. On the contrary, disasters are materialized risks. However, the perception of risk is not unanimous across a given society. Although there are not private perceptions of risk, as they all are influenced by cultural representations and social imaginaries, different social groups—be they defined by age, social class or ideological affiliation—will appraise differently what is a threat and how acceptable it is [46]. Some people are reckless, while others drink bleach. From this viewpoint, there is no such thing as an objective risk, but sociocultural contexts in which the latter are constructed and perceived.

Risks are not just social constructions, though; as the COVID-19 comes to show, they refer to particular realities that can be evaluated in an objective manner. Rationalistic approaches attempt to do just that—they seek to distinguish between reality and perception. More to the point, they try to locate particular risks and to identify their causes, building predictive models and designing pre-emptive policies with the help of experts [47]. Their usefulness has been proven during the first wave of the COVID-19 pandemic, as the effective responses of countries such as Germany or Austria demonstrate. The public response to a public health challenge, after all, cannot depend on shared beliefs. It is debatable whether a risk can be objectively measured, but perhaps the question is whether the public perception of risk can be aligned with their expert appraisal. Ultimately, Beck is right when he claims that acceptable risks are those that are accepted [9]. While some people drive a car, they never fly; others reject nuclear energy but eat processed meat. For each of those risks, different people behave differently. Social processes through which a threat is perceived, categorized and evaluated thus remain decisive for understanding how particular risks are socially met. Risks are never unmediated, but perceived—a perception that results from a process of social construction in which cultural, psychological and social factors play a key role [48]. Media coverage is also essential, exerting a strong influence on our sense of vulnerability: plane crashes are breaking news, whereas car accidents are barely reported [49].

In this regard, the COVID-19 pandemic offers some remarkable insights. On the one hand, epidemiologists have long warned that a greater medical knowledge and a more intense monitoring of infectious diseases may create new fears by making individuals aware of previously overlooked threats [8]. Yet the coronavirus pandemic has shown that epidemic risks can also be fatally underestimated; the slow reaction of some governments after the outbreak, notably the Spanish and the British ones, facilitated the spread of the virus in their countries. National agendas can hinder public action—the political interest of the government in celebrating the Women's Day demonstration on March 8th led the Spanish authorities to downplay the threat, misleading the public as far as the lethality of the disease was concerned. The slogan that captured this view is eloquent enough: *Solo es una gripe* (it's just a flu). Once the virus started to send people to the hospital, the imagined likeness to flu was lost and, in view of its menacing features, the public was ready to accept restrictive measures [50] (see Kasperson et al. 1998). A new virus seems thus liable to cause a strong disjunction between the accepted risk and the real danger—an imagined threat is thus accepted, while the real one is ignored.

Such misreading, which is more likely to occur in the face of an emerging virus, was reinforced in those countries whose governments reacted either too slowly or just overconfidently, thus leading to unfortunate political decisions about how best to respond to the pandemic. Something of the sort happened during the first years of the AIDS epidemic, before scientists had a firm grasp of the disease's nature. A part of the gay community, among them philosopher Michel Foucault, openly questioned the epidemiological response to the virus, which they took as an instrument of social control maliciously directed against a minority [51]. Risk evaluation is rarely unanimous, as the social construction of risk cannot escape the dynamics of democratic pluralism—nuclear energy is a case in point. In the case of COVID-19, there is no single answer to the question of how to handle the trade-off between an epidemiologically-led lockdown and the halting of economic activity. Political conflicts and moral debates have a lot to with risk evaluation. In a hyper technological society in which new dangers emerge, their centrality can only increase.

However, disagreements on the scale and significance of social risks do not only arise from the lack of a rational consensus about them, but also from the way in which the subjective experience of anxiety is anchored to our categories of knowledge [25]. In other words, the social assimilation of both imagined risks and realized disasters—such as the coronavirus pandemic—can barely be understood without taking emotions and moods into account—fear, panic, recklessness, hope. This affective response is influenced by political messages, but they have a life of their own and they largely determine how citizens react to particular threats. Whereas angry individuals sticks to their beliefs and are not willing to compromise, frightened ones will accept more easily those public decisions that are presented as necessary for guaranteeing their safety [52]. From the viewpoint of the public authority, then, it is better to frighten people than to make them mad. In the face of a potentially lethal outbreak, people who are angry demand a public response from their representatives and hold them accountable, whereas fear leads to the adoption of precautionary measures such as washing hands or wearing face masks [53]. Thus, emotions count, but it is not easy to know which of them are going to prevail in a given situation.

Unlike the realistic approach, which tries to evaluate threats from a rational standpoint, and the culturalist focus on the context in which risks are perceived, the post-modern perspective claims that *threats themselves* are socially constructed [54]. To put it differently, it is not just risk that is imagined—dangers too. Inspired by a conventional reading of Foucault, sociologist Mitchell Dean has explained this stance in the following way [55]:

> There is no such thing as a real in reality. Risk is a way—or rather, a set of different ways—of ordering reality, of rendering it into a calculable form. It is a way of representing events that makes them governable in particular ways, with particular techniques, and for particular goals. It is a component of diverse forms of calculative rationality for governing the conduct of individuals, collectivities and populations.

From this standpoint, the idea that a risk can be figured out is preposterous. The meaning of risk does not lie in its objective dangerousness, but in the governing regime that it helps to implement, as well as in the moral technologies that it brings about. This suspicion explains Agamben's interpretation of the virus as a political strategy intended to declare a biopolitical state of exception [45]. It can certainly not be disregarded that elements of the exception may remain in place after the outbreak is over, thus eroding liberal democracies. However, the imagined risk has actually become a disaster—zoonotic spillover is no longer an imagined abstraction but a threat that has caused more than a million deaths across the world. Even if we accept the idea that objective threats do not exist, it should be accepted that all risks are measurable through a number of rational operations that seek to determine how likely it is that a given risk finally occurs. For instance, we know that a strong hurricane has more chances of reaching the coast of Florida than an asteroid of hitting Earth—and so on and so forth. Such calculations cannot produce absolute truths, but they help to guide public and private decisions in the face of uncertainty.

However, the post-modern approach to risk is in itself one of the ways in which Western cultures respond to risk. Ultimately, it is a manifestation of critical reason—a suspicion about the aims of those actors who identify risks and derive normative consequences from them. It can be a well-grounded suspicion: homosexuality was persecuted in the past, and still is in some countries, as an unhealthy practice. Being alert to the political manipulation of public health emergencies makes sense, provided that this alertness does not prevent the recognition that many threats are painfully real. A thin line separates a sophisticated mistrust from an unfounded suspicion, as the number of conspiracy theories about the COVID-19 pandemic that have been put into circulation by populistic groups or plain amateurs demonstrates.

6. COVID-19 and Its Metaphors

Social theory offers different approaches to risk, but this does not entail the need to choose among them. On the contrary, they all provide valuable insights—rationalists objectively assess the probability and dangerousness of particular threats, while culturalists help us to understand why the latter are assimilated in different ways by different cultures and social groups, including the key role played by the media, and post-modernists alert us to the chance that risks are turned into elements of a given regime of government. Those approaches that focus on individual subjectivity make valuable contributions as well, be they phenomenologists that study the meanings that risks are attributed in different contexts, or psychoanalytical theorists who try to decipher the unconscious dynamics at play. Let us think of the massive purchase of food and toilet paper in the early moments of lockdowns, which can be interpreted respectively as enacting the morbid phantasy of starvation and as expressing a need for cleanliness. Likewise, risk can also become a transgressive force, less a danger than a practice from which strong emotions can be derived—clandestine parties in Berlin or London during lockdowns suggest that much. Thus, the social reaction to the coronavirus pandemic can be contemplated from several vantage points, and all of them are useful in their own ways.

Cultures adapt to risk, then, as they learn to make calculations or to produce obligations. In countries that suffered the SARS outbreak back in 2002–2004, such as Taiwan and Singapore, past experience has shaped an efficient response to COVID-19 [56]. By contrast, Western societies for which zoonotic spillovers are a less familiar affair reacted more slowly and less efficiently. Yet the coronavirus pandemic has taken place in the context of a global culture—one that does not replace local and national ones, but which is intertwined with them and thrives on hybridization [57]. Strictly speaking, the novelty does not lie in the global reach of the pandemic, since the Spanish Flu was also a world disease that emerged at the end of a long period of mercantilist globalization. The latter is, however, more intense today. Most decisively, digital technologies of communication have substantially changed how people see their societies in relation to the world, and vice versa. In the global public sphere, the dimensions of which are still modest, educated citizens from all over the world exchange information, express their opinions and communicate their emotional states [58]. For better or worse, this trend has been reinforced by the global pandemic.

Mostly, it has been for the good. Despite the fear of a xenophobic reaction against Chinese residents abroad during the first months of the epidemic outbreak, this has not happened or has been almost unnoticeable. There have been racist episodes and some countries have seen how negative feelings against particular minorities increased [59]. So far, however, no massive or violent reaction against social groups on account of their supposed role in creating or extending the disease has been registered—national populistic movements have not succeeded in implementing their exclusionary agenda. This marks a stark contrast with the AIDS outbreak in the early 80s, when the idea that described the disease as a moral punishment against practitioners of deviant forms of sexuality was widespread [60]. The notion of the abject, understood as a source of horror and fascination that destabilizes our psyche in the presence of an outer threat [61], has been reserved in the case of COVID-19 for certain Asian cultural practices related to the source of the zoonotic spillover: the image of the bat soup has figured prominently in the Western imaginary of the pandemic [62]. This fixation does not correspond to a wave of ethnical hostility, though. A triumph of the emerging global culture can be cautiously announced as far as the COVID-19 pandemic is concerned.

To test this hypothesis, the sociological concept of disease identity can be useful. According to it, every major disease is associated to a particular identity, which takes shape once the illness acquires public visibility for the first time and tends to keep stable in the future—in a manner that is akin to the stickiness of racial and ethnic stereotypes [63]. It is a collective construction that attributes specific features to a given disease and its carriers, and which originates in a discursive process in which scientists, political leaders, journalists and (increasingly) citizens take part. Such identities are not without consequences, as they influence public policies and private attitudes: they make up the dominant image of the disease. In her classic study on tuberculosis, Sontag described how the imaginary surrounding this disease portrayed it as a romantic condition, associated to passionate people gifted with an artistic sensitivity [64]. On its part, the negative effects of the moral stigmatization suffered by patients of AIDS have been richly documented [65].

But which is the emerging identity of COVID-19? At this point, we can only speculate. Despite president Trump's attempts to label it as the Chinese virus, the quick spread of the disease across the world soon weakened the strength of such symbolic association. To be sure, the source of this zoonotic spillover is not meaningless, as Chinese society goes through a wild stage of modernization that facilitates epidemic outbreaks. Yet the labeling has not become popular and will hardly help us to define the disease's identity. COVID-19 stands out as a highly infectious disease, and that is the reason why it has spread so quickly around the globe, potentially threatening all its inhabitants—although it is more lethal for the elderly and for those who suffer other medical conditions, anyone can develop the disease, and there is no particular minority or ethnic group that finds itself at greater risk than others. That is why COVID-19 rather emerges as a universal malady that, attending to the historical context in which appears, might be described as the first disease of the Anthropocene. Never mind that this association is not straightforward, as we have seen above, since what counts for fixing a disease's identity is collective perception—be it right or wrong.

Moreover, a typical symptom of the worst cases of COVID-19 is acute respiratory failure, and this severe condition can be symbolically related to the undesirable consequences of industrialization, the more worrying expression of which is climate change. French president Emmanuel Macron has made this tenuous connection explicit, by saying that fear of dying by suffocation reflects the fear of breathing the polluted air of our cities [66]. Small wonder, then, that face masks have become the universal symbol of this pandemic—despite them being already around during the Spanish Flu. Yet the mask is an ambivalent sign. On the one hand, it points to a partial suspension of individual autonomy, thus representing an emancipatory regression. On the other, it summarizes a communitarian feeling that equalizes us all despite the cultural differences that can be found once their use within particular contexts is analyzed. Some anthropologists foresee a normalization of the face mask due to the intensification of climate change, and as the air quality of our cities further deteriorates [67]. It could as well be forgotten, though, once wearing masks ceases to be compulsory. The symbolic association between the virus and the acceleration of global modernity seems, however, established. Hence the successful metaphor that displaces the subject of the disease: instead of the virus being the one that kills, human beings have been presented as the virus that devastates the planet and suffer the consequences of their reckless behavior in the form of a zoonotic disease.

7. Rethinking Global Risks after Corona

It is tempting to reach the conclusion that COVID-19 confirms that we live in a global risk society in which latent threats turn into real disasters at worrying speed. The planet has become a claustrophobic place, and there is some irony to this: globalization has made the planet smaller. Yet we should not jump to conclusions that quick. It is not surprising that risk society theory resonated so strongly when it was first formulated—the original German edition coincided with the Chernobyl disaster. Western societies, though, when observed from a different angle, can lead to different conclusions about the prevalence of risk.

The category of risk, as we have seen, emerges with modernity. Prior to that, human endeavors are seen as mediated by danger and misfortune. It is precisely because humans increase their ability to shape their environment and learn how to relate to it that the notion of risk appears—as a contingency that escapes from calculation, an occasional system failure that leads to an accident. Through technological development and bureaucratic systems, complexity and randomness are reduced at the expense of creating new threats that are inherent to this societal transformation. As it happens, modern societies are reasonably efficient in exerting control over such threats, which needless to say cannot be totally eradicated. It suffices to think of the lethality that accompanied Black Death or the Spanish Flu, as opposed to the moderate lethality that characterizes COVID-19. From this viewpoint, the risk society might rather be contemplated as the control society as far as the ability to minimize threats is concerned. Western societies have grown in complexity in the last three centuries, but this process has not been accompanied by a proportional rise in the number of disasters—although disasters, of course, do exist. The fact that disasters capture media attention, while prevented risks do not, helps to explain this paradox of modernity. The same applies to social expectations, which have changed significantly in the last three centuries: the normality of disaster has given way to the perception that disasters are exceptional.

Let us take into consideration the century in which health policy was seriously developed for the first time, namely the 19th century. It was then that, as historian Jürgen Osterhammel points out, the democratization of the long life took place. What makes it possible for humans to live longer lives is the increase in material prosperity, a process the first positive outcomes of which were visible in Europe between 1890 and 1920, and only later in other regions of the world [68]. Yet this would have never been achieved without the emerging knowledge on disease prevention and the invention of public healthcare, which has since then turned into a state duty. By improving water quality, for instance, cholera's spread was contained. Early modernity was indeed insalubrious, and the biological cost of modernization was higher for those countries that industrialized first. At the same time, the universalization of the new health values was not always accompanied by the provision of the means necessary for realizing them, a political limitation that hindered the rise of the *homo hygienicus* [68]. From then on, the human being becomes the animal that successfully disinfects itself.

In the 19th century, then, an epidemiological transition begins that will run parallel to the demographic shift—the substantial reduction of mortality is mostly due to the reduced likeliness of dying from infection. Following Osterhammel, it was characteristic of the time that disease was more easily communicated and yet also more successfully treated. Plague and typhus were weakened between 1600 and 1750. Then it was the turn of scarlet fever, diphtheria and whooping cough, and in the 19th century respiratory diseases lost strength, with the exception of tuberculosis. New diseases appeared, of course: meningitis was discovered in 1805. Still, it can be safely affirmed that the 20th century established a new medical regime thanks to the coming together of Koch and Pasteur's insights, the end of smallpox (it was eradicated in 1980), and the salutary influence of the progressive social hygiene movement. Sometimes, though, some luck was required: the transition to stone architecture reduced the natural habitat of domestic rats that carried the plague.

This historical account offers a number of lessons regarding pandemics and risk. The first, most evidently, is that the human animal remains constantly threatened by biological accidents—no matter how many diseases have been eradicated, others will follow. Yet the second is that this very human animal has developed a remarkable ability to prevent them, with the proviso that such capacity is initially less effective when we are confronted with new viruses. Epidemic risks, in sum, cannot be eliminated. Furthermore, even though they can turn into disasters, thus completing the shift from imagined risk to real catastrophe, major epidemic outbreaks are not that common. The COVID-19 pandemic is a system failure rather than an expression of a failed system. A failed system would be one in which pandemics are common and lethal; a system failure describes an event in which well-known precautionary measures, regarding for instance food security, are not adopted. Chinese wet markets, widely believed to be at the source of both the COVID-19 and the SARS outbreaks, are clear cases of risk misgovernment. The lack of food security is the key risk factor as far as zoonotic spillovers are concerned. As the journalist James Palmer puts it, the important thing is not *what* is eaten but *how* it is eaten, a distinction that the Chinese society does not feel compelled to make [69]:

> The country's food safety standards are notoriously bad, despite numerous government-led initiatives to improve them. Food scandals are common, and diarrhea and food poisoning are a distressingly regular experience. Markets, like Huanan, that aren't licensed for live species nevertheless sell them. Workers are undertrained in basic hygiene techniques like glove-wearing and hand-washing.

Climate change may very well bring about new pathogens humans will have to defend themselves against. However, if modernity had never occurred, if a new healthcare strategy based on disease prevention and personal hygiene had never been implemented, how many highly lethal pandemics might have ravaged human societies? On the contrary, those societies have learnt to prevent and cope with epidemic outbreaks—risk control has been reasonably efficient. That is why disasters are perceived as failures. Yet the yardstick according to which risk management should be evaluated cannot be the elimination of all disasters, a utopian goal that can only be sustained by a naive belief in the power of reason.

It is worth insisting on this point: risk is consubstantial to human activity. The right not to be subject to threats that people have not consented to cannot be granted, because social inaction would follow; almost nothing could be done lest risk emerge [70]. Correspondingly, total safety does not exist. In other words, a *decision* will always involve an unavoidable risk and the same applies to the *lack of decision*, which is in itself a decision. According to Luhmann, so many things can go wrong that rational calculation becomes very difficult—the key question being upon who or what the acceptance or rejection of a risk depends. He presents his point in the following way [71]:

> what can occur in the future also depends on decisions to be made at present. For we can speak of risk only if we can identify a decision without which the loss could not have occurred. (...) For the concept as we intend to define it, the only requirement is that the contingent loss be itself caused as a contingency, that is to say that it be avoidable.

Therefore, a risk can be attributed to a decision, while a danger comes from the outside. In order to know whether something is a risk or a danger, Luhmann warns, observers must be observed, since they are the ones who decide about the source of the threat. Thus an unknown risk, one that has never been identified or labeled, does not exist—it can only appear, out of the blue, as a danger [10]. What, though, about COVID-19? Was it a risk or a danger? It was, most obviously, a risk; the chance that a new coronavirus could spread globally was almost taken for granted after the outbreak of SARS in the early 20th century. The reach of the pandemic was not anticipated, nor was prevention enough to stop it. However, the ability to identify risks in a complex global society is not in harmony with the ability to minimize the likeliness of their realization—risks abound, and resources, including public attention and political decisiveness, are scarce. Yet this imbalance can be attributed to political reasons, as the stark contrast between the performances of different societies in the fight against the pandemic comes to show: those who have displayed a more efficient strategy have made use of massive testing and tracking devices, as opposed to those who could not properly identify the sources of infection, and thus resorted to banning social activity to prevent the spread of the virus. This is but another manifestation of the modernity shortage described in this paper.

Admittedly, the social and ecological changes that are characteristic of the Anthropocene may end up demanding a completely new approach to epidemic risks—pandemics may become the new normal [71]. In that case, which remains to be seen, there is no alternative to accepting that total safety is impossible, and the best must be done for preventing this old natural threat. This entails, reinforcing food safety and educating the public about the dangerousness of cultural practices that put us in intimate contact with wild animals. Science is not enough, of course, as it cannot decide alone which risks are socially acceptable—experts are just one of the voices that are heard in the *polis* [72]. Likewise, reinforcing the role of experts does not lead to the elimination of risks. As Luhmann suggests, it cannot be a coincidence that risk theory develops in parallel with scientific specialization; the more we know, the better we realize that there is still much that we do not know [71].

A crucial irony of risk management comes to the fore: the *realized* danger produces far more emotional impact on the population than the *averted* risk. The latter is invisible, while the other captures our attention and dominates the public conversation. As a result, a successful risk management leads to failure—only disasters make it to the headlines. This suggests that perceiving contemporary society as a risk society, a view that will surely be reinforced by the pandemic, is a misperception that fails to consider all the potential risks that do not turn into disasters. Instead of emphasizing the ability of modern societies to control their relationships with the environment, there is a focus on the negative aspect of that exchange. A culture of fear is thus fostered which, in the case of pandemics, adopts elements from the forbidden knowledge narrative—a virus is hidden in the jungle in which human beings dare to penetrate. The social relation with risk is described in negative terms: the latter escapes human control, and while we have the power to destroy whole environments and livelihoods, there is not much that we can do about the dangers that threaten our daily existence [73].

It may well be that the psychological bias that leads us to exaggerate the likeliness and deadliness of potential risks, coupled with the sensationalist media handling of those that turn into disasters, serves a means to prevent them. On the other hand, modernity has not suppressed risks; on the contrary, it has enlarged their variety and range. Yet labeling late modern societies as risk societies is, in view of the latter's ability to handle complexity, debatable. Again, the problem lies in the yardstick used for appraising risk management: the utopia of a riskless society is not a realistic one and the question remains open—it is a political question—as to how many risks, and which ones, are socially acceptable. Still, as COVID-19 shows, major disasters can happen, and actually great epidemic outbreaks had been common until recently—the Spanish Flu, polio outbreaks, AIDS. A moderate approach to risk in modernity barely helps when a risk turns into a disaster. In fact, their occurrence is received with perplexity as people in the Western world can hardly believe that a virus can kill so many people and upend their societies so quickly. However, yes, a virus can do this—especially when prevention is weak, and governance fails. Yet this is no reason for discarding a nuanced approach to risks, a multi-faceted subject that is best understood through the combination of different perspectives.

8. Conclusions

The COVID-19 pandemic is a socionatural threat that reproduces a well-known danger, that of zoonotic infectious disease. This kind of virus belongs to the category of crowd diseases, and can be associated to human sedentariness rather than to late-modern conditions *per se*. Corona, however, appears in the context of late modernity, and thus it reproduces some of its features—above all, the speed at which the virus has spread globally in a globalized world. Likewise, this particular virus has succeeded in jumping onto humans in the Anthropocene, and this might lead to the claim that COVID-19 is the first disease of this new epoch. Literally, it probably is. However, this pandemic shares some key features with the Spanish Flu and other past outbreaks, including the global epidemic of bubonic plague known as the Black Death that lasted from 1347 to 1351. If the virus had actually originated in the meat industry, the Anthropocene framework would be more apposite for explaining its appearance and diffusion. It seems certain, though, that the source is a Chinese wet market, where the relation between humans and animals is anything but modern. The pandemic is thus rather a Holocene risk that remains in the Anthropocene. That said, the latter's conditions make zoonotic spillovers more common due to the human colonization of wet and tropical habitats in which dangerous pathogens abound. Social sciences should thus provide a nuanced view of epidemic risks, including the COVID-19 outbreak, taking the most obvious theoretical associations with a grain of salt and making others, perhaps less conspicuous ones, more explicit.

In this regard, I have argued that it would be a mistake to frame the COVID-19 pandemic as a negative side-effect of modernity. As such, zoonotic infections are a pre-modern threat that have survived the advent of modernity, despite the success of the new hygienic and immunological policies implemented in the late 19th and the early 20th century. Given the number of viruses and bacteria that populate the Earth, eradicating zoonotic infections does not seem feasible. Human societies should then focus on minimizing this kind of risk by increasing their association to modern immunological strategies, and at the same time by refining their relation to natural environments. While zoonotic infections have lost virulence, they seem to occur more often due to human penetration into wild habitats and trafficking with wild animals. Needless to say, the industrial food system should also be criticized, reformed and/or reduced on both health and moral grounds. The emergence of new viruses due to global warming is also a concern. Yet the COVID-19 pandemic is to be explained in a different way—getting it right is a first step toward preventing the next global pandemic.

Funding: This text is part of the project Antropoceno: Sostenibilidad y democracia en el nuevo contexto planetario, funded by the Universidad de Málaga.

Acknowledgments: I would like to thank to the Editor and Reviewers for their comments and suggestions.

Conflicts of Interest: The author declare no conflict of interest.

References

1. Cheng, V.; Lau, S.; Woo, P.; Yuen, K. Severe acute respiratory syndrome coronavirus as agent of emerging and reemerging infection. *Clin. Microbiol. Rev.* **2007**, *20*, 660–694. [CrossRef]
2. Menachery, V.D.; Yount, B.L., Jr.; Debbink, K.; Agnihothram, S.; Gralinski, L.E.; Plante, J.A.; Graham, R.L.; Scobey, T.; Ge, X.-Y.; Donaldson, E.F.; et al. A SARS-like cluster of circulating bat coronavirus shows potential for human emergence. *Nat. Med.* **2015**, *21*, 1508–1513. [CrossRef] [PubMed]
3. Rasul, I. The Economics of Viral Outbreaks. *Pap. Proc. Am. Econ. Assoc.* **2020**, *110*, 265–268. [CrossRef]
4. Spinney, L. *Pale Rider: The Spanish Flu of 1918 and How It Changed the World*; Jonathan Cape: London, UK, 2017.
5. Tabarrok, A. The Forgotten 1957 Pandemic and Recession, Marginal Revolution, 24 March 2020. Available online: https://marginalrevolution.com/marginalrevolution/2020/03/the-forgotten-1957-pandemic-and-recession.html (accessed on 26 November 2020).
6. Lorrain, F.G. Ces Pandémies de Grippe que la France a Oubliées, Le Point, 4 April 2020. Available online: https://www.lepoint.fr/sante/ces-pandemies-de-grippe-que-la-france-a-oubliees-04-04-2020-2370065_40.php (accessed on 26 November 2020).
7. Quanmen, D. *Spillover: Animal Infections and the Next Human Pandemic*; Bodley Head: London, UK, 2012.
8. Honigsbaum, M. *The Pandemic Century. One Hundred Years of Panic, Histeria, and Hubris*; W.W. Norton, W.W., & Company: New York, NY, USA, 2019.
9. Beck, U. *World at Risk*; Polity: Cambridge, UK; Malden, MA, USA, 2009.
10. Luhmann, N. *Risk: A Sociological Theory*; Aldine de Gruyter: New York, NY, USA, 1993.
11. Crosby, A. *The Columbian Exchange: Biological and Cultural Consequences of 1492*; Praeger: Westport, CT, USA, 2003.
12. McNeill, W. *Plagues and Peoples*, 3rd ed.; Anchor Books: New York, NY, USA, 1998.

13. Diamond, J. *Guns, Germs & Steel: A Short History of Everybidy for the Last 13000 Years*; Vintage: London, UK, 1998.
14. Wilson, E. *The Diversity of Life*; Penguin: London, UK, 1993.
15. Zalasiewicz, J.; Waters, C.N.; Wolfe, A.P.; Barnosky, A.D.; Cearreta, A.; Edgeworth, M.; Ellis, E.C.; Fairchild, I.J.; Gradstein, F.M.; Grinevald, J.; et al. Making the case for a formal anthropocene epoch. An analysis of ongoing critiques. *Newsl. Stratigr.* **2017**, *50*, 205–226. [CrossRef]
16. Hamilton, C. *Defiant Earth: The Fate of Humans in the Anthropocene*; Polity: Cambridge, UK, 2017.
17. Liu, Y.; Gayle, A.; Wilder-Smith, A.; Rocklöv, J. The reproductive number of COVID-19 is higher compared to SARS coronavirus. *J. Travel Med.* **2020**, *27*. [CrossRef] [PubMed]
18. Virilio, P. *The Original Accident*; Polity: Cambridge, UK; Malden, MA, USA, 2006.
19. Malm, A. *Corona, Climate, Chronic Emergency: War Communism in the Twenty-First Century*; Verso: London, UK; New York, NY, USA, 2020.
20. Sloterdijk, P. *In the World Interior of Capital: For a Philosophical Theory of Globalization*; Polity: Cambridge, UK; Malden, MA, USA, 2013.
21. Schmitt, C. *Land and Sea*; Plutarch Press: Washington, DC, USA, 1997.
22. Schmitt, C. *The Nomos of the Earth*; Telos Press: New York, NY, USA, 2003.
23. Hamilton, C.; Adolphs, S.; Nerlich, B. The meanings of "risk". A view from corpus linguistics. *Discourse Soc.* **2007**, *18*, 163–181. [CrossRef]
24. Hacking, I. *The Taming of Chance*; Cambridge University Press: Cambridge, UK, 1990.
25. Wilkinson, I. *Anxiety in a Risk Society*; Sage: London, UK, 2001.
26. Beck, U. *Risikogesellschaft: Auf Dem Weg in Eine Andere Moderne*; Suhrkamp: Frankfurt, Germany, 1986.
27. Vieten, U.; Eranti, V.; Blokker, P. Thinking and Writing in the Time of Pandemic COVID-19. *Eur. J. Cult. Political Sociol.* **2020**, *7*, 117–122. [CrossRef]
28. Giddens, A. *The Consequences of Modernity*; Polity: Cambridge, UK, 1990.
29. Giddens, A.; Lash, S. *Reflexive Modernization. Politics, Tradition, and Aesthetics in the Modern Social Order*; Polity Press: Cambridge, UK, 1994.
30. Beck, U. *Gegengifte: Die organisierte Unverantwortlichkeit*; Suhrkamp: Frankfurt, Germany, 1998.
31. Beck, U. World Risk Society and Manufactured Uncertainties. *IRIS* **2009**, *I*, 291–299.
32. Giddens, A. *Modernity and Self-Identity*; Polity: Cambridge, UK, 1991.
33. Patz, J.A.; Confalonieri, U.E.; Amerasinghe, F.P.; Chua, K.B.; Daszak, P.; Hyatt, A.D.; Vasconcelos, P.; Mahamat, H.; Mutero, C.; Whiteman, C.; et al. Human health: Ecosystem regulation of infectious diseases. In *Ecosystems and Human Well-Being: Current State and Trends Volume 1*; Condition and Trends Working Group, Ed.; Island Press: Washington, DC, USA, 2005; pp. 391–415.
34. O'Callaghan-Gordo, C.; Antó, J. COVID-19: The disease of the Anthropocene. *Environ. Res.* **2020**, *187*, 109683. [CrossRef]
35. Rosol, C.; Nelson, S.; Renn, J. Introduction: In the machine room of the Anthropocene. *Anthr. Rev.* **2017**, *4*, 2–8. [CrossRef]
36. Dyer-Witheford, N. Struggles in the planet factory: Class composition and global warming. In *Interrogating the Anthropocene*; Jagodzinski, J., Ed.; Springer International Publishing: Cham, Switzerland, 2018; pp. 75–103.
37. Spaargaren, G.; Mol, A.P.J.; Buttel, F.H. (Eds.) Introduction: Globalization, modernity and the environment. In *Environment and Global Modernity*; Sage: London, UK, 2000; pp. 1–16.
38. Ferreira, C.M.; Sa, M.J.; Garrucho Martins, J.; Serpa, S. The COVID-19 contagion-pandemic dyad: A view from social sciences. *Societies* **2020**, *10*, 77. [CrossRef]
39. Kossin, J.P.; Knapp, K.R.; Olander, T.L.; Velden, C.S. Global increase in major tropical cyclone exceedance probability over the past four decades. *Proc. Natl. Acad. Sci. USA* **2020**, *117*, 11975–11980. [CrossRef] [PubMed]
40. Casadevall, A.; Kontoyiannis, D.; Robert, V. On the emergence of *Candida auris*: Climate change, swamps, and birds. *mBio* **2019**, *10*, e01397-19. [CrossRef] [PubMed]
41. Hirschfeld, K. Microbial insurgency: Theorizing global health in the Anthropocene. *Anthr. Rev.* **2020**, *7*, 3–18. [CrossRef]
42. Mills, J.; Gage, K.; Khan, A. Potential influence of climate change on vector-borne and zoonotic diseases: A review and proposed research plan. *Environ. Health Perspect.* **2010**, *118*, 1507–1514. [CrossRef] [PubMed]

43. Keys, P.W.; Galaz, V.; Dyer, M.; Matthews, N.; Folke, C.; Nyström, M.; Cornell, S.E. Anthropocene risk. *Nat. Sustain.* **2019**, *2*, 667–673. [CrossRef]
44. Mythen, G. *Ulrich Beck: A Critical Introduction to the Risk Society*; Pluto Press: London, UK, 2004.
45. Agamben, G. Lo Stato D'eccezione Provocato da Un'emergenza Immotivata, Il Manifesto, 26 February 2020. Available online: https://ilmanifesto.it/lo-stato-deccezione-provocato-da-unemergenza-immotivata/ (accessed on 26 November 2020).
46. Douglas, M. *Risk and Blame: Essays in Cultural Theory*; Routledge: London, UK, 1992.
47. Lupton, D. (Ed.) *Risk and Sociocultural Theory: New Directions and Perspectives*; Cambridge University Press: Cambridge, UK, 1999.
48. Slovic, P. *The Perception of Risk*; Earthscan: London, UK, 2000.
49. Lübbe, H. Security. Risk Perception in the Civilization Process. In *Risk is a Construct. Perceptions of Risk Perception*; Rück, B., Ed.; Knesebeck: Munich, Germany, 1993; pp. 23–40.
50. Kasperson, K.; Renn, O.; Slovic, P.; Brown, P. The social amplification of risk: A conceptual framework. In *Risk and Modern Society*; Löfstedt, R., Frewer, L., Eds.; Earthscan: London, UK, 1998; pp. 149–162.
51. Miller, J. *The Passion of Michel Foucault*; Doubleday: New York, NY, USA, 1994.
52. Brader, T.; Marcus, G. Emotion and Political Psychology. In *The Oxford Handbook of Political Psychology*; Huddy, D.L., y Jack Levy, S., Eds.; Oxford University Press: Oxford, UK, 2013; pp. 165–205.
53. Brader, T.; Gorenendyk, E.; Valentino, N. Fight or flight? When political threats arouse public anger and fear. In Proceedings of the Annual Meeting of the Midwest Political Science Association, Chicago, IL, USA, 31 March–3 April 2011.
54. Lupton, D. (Ed.) Postmodern reflections on "risk", "hazards" and life choices. In *Risk and Sociocultural Theory: New Directions and Perspectives*; Cambridge University Press: Cambridge, UK, 1999; pp. 12–33.
55. Dean, M. Risk, calculable and incalculable. *Soz. Welt* **1998**, *49*, 25–42.
56. Keck, F. Asian tigers and the Chinese dragon: Competition and collaboration between sentinels of pandemics from SARS to COVID-19. *Centaurus* **2020**, *62*, 311–320. [CrossRef]
57. Pietersee, J.N. *Globalization & Culture: Global Mélange*; Rowan & Littlefield: London, UK, 2019.
58. Sparks, C. The internet and the global public sphere. In *Mediated Politics: Communication in the Future of Democracy*; Bennett, L., Entman, R., Eds.; Cambridge University Press: Cambridge, UK, 2012; pp. 75–96.
59. Reny, T.; Barretto, M. Xenophobia in the time of pandemic. Othering, anti-Asian attitudes, and COVID-19. *Politics Groups Identities* **2020**. [CrossRef]
60. Williamson, J. Every virus tells a story. The meanings of HIV and AIDS. In *Taking Liberties. AIDS and Cultural Politics*; Carter, E., Watney, S., Eds.; Serpent's Tail: London, UK, 1989; pp. 69–80.
61. Kristeva, J. *Powers of Horror: An Essay on Abjection*; Columbia University Press: New York, NY, USA, 1982.
62. Darrach, A. The New Coronavirus and Racist Tropes, Columbia Journalism Review, 25 February 2020. Available online: https://www.cjr.org/analysis/covid-19-racism-china.php (accessed on 26 November 2020).
63. Taylor, R. The politics of securing borders and the identities of disease. *Sociol. Health Illn.* **2013**, *35*, 241–254. [CrossRef]
64. Sontag, S. *Illness as Metaphor & Aids and Its Metaphors*; Penguin: London, UK, 1990.
65. Tewksbury, R.; McGaughey, D. Stigmatization of persons with HIV disease. Perceptions, management, and consequences of AIDS. *Sociol. Spectr.* **1997**, *17*, 49–70. [CrossRef]
66. Financial Times. Interview with Emmanuel Macron, Financial Times, 18–19 April 2020. Available online: https://www.ft.com/content/3ea8d790-7fd1-11ea-8fdb-7ec06edeef84 (accessed on 26 November 2020).
67. Lasco, G. Why Face Masks Are Going Viral, Sapiens, 7th February 2020. Available online: https://www.sapiens.org/culture/coronavirus-mask/ (accessed on 26 November 2020).
68. Osterhammel, J. *The Transformation of the World: A Global History of the Nineteenth Century*; Princeton University Press: Princeton, NJ, USA, 2014.
69. Palmer, J. Don't Blame Bat Soup for the Coronavirus, Foreign Policy, 27th January 2020. Available online: https://foreignpolicy.com/2020/01/27/coronavirus-covid19-dont-blame-bat-soup-for-the-virus/ (accessed on 26 November 2020).
70. Teuber, A. Justifying risk. *Daedalus J. Am. Acad. Arts Sci.* **1990**, *119*, 235–254.
71. Stephen, C. Rethinking pandemic preparedness in the Anthropocene. *Healthc. Manag. Forum* **2020**, *33*, 153–157. [CrossRef] [PubMed]

72. Vegetti, M. *Chi comanda nella città*; Carocci: Roma, Italy, 2017.
73. Furedi, F. *Culture of Fear. Risk-Taking and the Morality of Low Expectations*; Cassell: London, UK; Washington, DC, USA, 1997.

Publisher's Note: MDPI stays neutral with regard to jurisdictional claims in published maps and institutional affiliations.

 © 2020 by the author. Licensee MDPI, Basel, Switzerland. This article is an open access article distributed under the terms and conditions of the Creative Commons Attribution (CC BY) license (http://creativecommons.org/licenses/by/4.0/).

Concept Paper

The COVID-19 Contagion–Pandemic Dyad: A View from Social Sciences

Carlos Miguel Ferreira [1,2], Maria José Sá [3], José Garrucho Martins [1] and Sandro Serpa [1,4,5,*]

[1] Interdisciplinary Centre of Social Sciences—CICS.NOVA, ISCTE—University Institute of Lisbon, 1649-026 Lisbon, Portugal; carlos.miguel.ferreira@iscte-iul.pt (C.M.F.); jj.garruchomartins@gmail.com (J.G.M.)
[2] Estoril Higher Institute for Tourism and Hotel Studies, 2765-273 Estoril, Portugal
[3] CIPES—Centre for Research in Higher Education Policies, 4450-137 Matosinhos, Portugal; mjsa@cipes.up.pt
[4] Department of Sociology, Faculty of Social and Human Sciences, University of the Azores, 9500-321 Ponta Delgada, Portugal
[5] Interdisciplinary Centre for Childhood and Adolescence—NICA, University of the Azores, 9500-321 Ponta Delgada, Portugal
* Correspondence: sandro.nf.serpa@uac.pt

Received: 30 July 2020; Accepted: 1 October 2020; Published: 6 October 2020

Abstract: The objective of this concept paper focuses on the relevance of the analytical potential of Social Sciences for understanding the multiple implications and challenges posed by the COVID-19 contagion–pandemic dyad. This pandemic is generating a global threat with a high number of deaths and infected individuals, triggering enormous pressure on health systems. Most countries have put in place a set of procedures based on social distancing, as well as (preventive) isolation from possible infected and transmitters of the disease. This crisis has profound implications and raises issues for which the contribution of Social Sciences does not seem to be sufficiently mobilised. The contribution of Social Sciences is paramount, in terms of their knowledge and skills, to the knowledge of these problematic realities and to act in an informed way on these crises. Social Sciences are a scientific project focused on interdisciplinarity, theoretical and methodological plurality. This discussion is developed from the systems of relationships between social phenomena in the coordinates of time and place, and in the socio-historical contexts in which they are integrated. A pandemic is a complex phenomenon as it is always a point of articulation between natural and social determinations. The space of the discourse on the COVID-19 pandemic can be understood as the expression of a coalition of discourses, i.e., the interaction of various discourses, combined in re-interpretative modalities of certain realities and social phenomena. The circumstantial coalitions of interests, which shape the different discursive records and actions produced by different agents of distinct social spaces, enable the acknowledgement and legitimation of this pandemic threat and danger, and the promotion of its public management.

Keywords: COVID-19; social sciences; inequality; pandemic; contagion; social distancing

---

1. Introduction

An epidemic of viral-appearing pneumonia of unknown aetiology emerged in Asia in December 2019. In January 2020, the identification of a new coronavirus was officially announced by Chinese health authorities and the World Health Organisation (WHO). It was first called NCoV 2019 and then SARS-CoV-2. This new virus is the agent responsible for the infectious respiratory disease called COVID-19. On 11 March 2020, WHO announced that COVID-19 could be qualified as a pandemic, the first one triggered by a coronavirus.

The epidemic phenomena are the expression of a violent clash between species [1] and between humans and non-humans [2,3]. Coronaviruses (CoV) entail a large group of viruses, some of which are pathogenic to humans, whose infections are usually associated with clinical respiratory manifestations, without high severity. Over the last decade, two new coronaviruses have emerged as highly pathogenic infectious agents for humans, causing potentially lethal infections. These coronaviruses have been responsible for the SARS (Severe Acute Respiratory Syndrome Coronavirus) and the MERS (Middle East Respiratory Syndrome-related Coronavirus) respiratory syndromes. Both SARS-CoV and MERS-CoV viruses have a zoonotic origin, i.e., their natural reservoir is animals, in particular some species of bats. The genesis of the SARS-Cov-2 coronavirus, although raising some doubts, also had a zoonotic origin [4].

In the process of naming the SARS-CoV-2 virus, WHO favoured a "politically correct" nomenclature, which sought to prevent the stigmatisation of the Chinese regions associated with the genesis of the pandemic. This potential stigmatisation was present in the designations "West Nile Virus", "Lhasa Fever" (a city in Nigeria, located in the Yedseram River Valley), "Ebola Virus" (a river in the Democratic Republic of Congo), "Middle East Respiratory Syndrome" and, in the H5N1 outbreaks, "from Fujian" (a province in South-East China) or "from Qinghai" (a lake in West China). The WHO's strategy to prevent this was to name viruses and viral outbreaks with an emphasis on the molecular structure. The focus on the molecular can be deemed as a way of blurring the relevance of economic and environmental factors in the emergence of new strains of a virus, particularly in the case of the H1N1 or H5N1 variants, i.e., industrial livestock. Since the late 1970s, with the globalisation of intensive pig and poultry farming, outbreaks of the increasingly virulent influenza have multiplied. In industrial animal husbandry, stocking density and genetic homogeneity provide "the perfect incubator" for viruses, which have an evolutionary interest in transmitting faster and becoming more virulent [5].

The COVID-19 pandemic is generating a global threat. By mid-July 2020, there were 14 million people infected, 7.7 million considered cured and about 600,000 dead. However, the exact number of those infected due to a lack of testing or those killed due to underreporting is not known. This pandemic has most significantly affected the older population and/or individuals with respiratory complications and other similar pre-existing diseases [6–8].

The vast majority of countries [9–11] have imposed a set of preventive procedures and devices, in addition to medical tests, based on isolation, quarantine, community containment and physical distancing [12,13], travel restrictions, hand hygiene and the banning of events and gatherings. These measures have led to the temporary closure of several economic and social institutions, a relative national lockdown in several countries and enormous pressure on health systems [12,14–23]. However, countries such as Sweden, South Korea and Taiwan did not impose total containment but followed a strategy that articulated quarantine targeting risk groups, monitoring, and large-scale testing, together with social distancing measures (even the use of masks).

The extraordinary and temporary measures adopted by a significant number of countries sought to keep people confined to their homes, in prophylactic isolation and, preferably, in a teleworking regime where the functions concerned were allowed. This situation could not take place if people were unable to telework or to carry out activities deemed essential and a priority. In some countries, individuals should remain in their homes, although they could go out for exercise once a day, to buy food and other essential items. The sick and elderly should also stay in their homes. Several restrictions were imposed on the interactions between people who were not cohabiting; for example, they could not interact with more than one person and should respect the distance of two meters from other people [10,24].

Without yet knowing the specificities of the effects of the different dimensions of the pandemic on mental health, it is possible, however, to assess its importance with the data already available. The stigmatisation and production of scapegoats—already known in other epidemics and other historical moments—in the case of COVID-19, have victimised some health professionals and individuals from minority social groups. The uncertainties and partial mistrust of medical knowledge and information, together with the publication of conspiracy theories have also added to the emergence of a diffuse living disease that does not facilitate mental health. In this context, some disturbances, which are characteristic of uncertainty/unpredictability and lack of self-control in the relationship with oneself and with others, have emerged. As indicators of the fragility of these relationships, the increase in generalised anxiety and obsessive-compulsive disorders, aggressive depressions, insomnia and feelings of frustration [25] are understandable.

These various measures have a profound impact on everyday life, and the future consequences for the reconfiguration of social life are unknown [20,26]. This is a dynamic situation with varying amplitude, taking into account the geographical, economic, societal and cultural contexts affected. Social distancing (this is the term medically consecrated and, therefore, used in this article, although the physical distance corresponds to what is under discussion) may be effective in preventing contagion and deaths, although it can be perceived as insidious to the economic activity since social distancing usually entails economic distancing: most industries require workers to develop close interactions to produce goods, while various services require close contact between customers, users and suppliers, or between customers [13].

However, inequality in various dimensions is a critical element, affecting responses to COVID-19 and even exacerbating inequalities [11,27–29]. Poorer populations, and especially older ones, are more vulnerable to infection and more vulnerable to the most serious consequences. The pandemic is generating a significant economic crisis and, consequently, unemployment rates will tend to increase substantially. Furthermore, weakened social safety nets further threaten the health and social security of the most vulnerable social categories also in day-to-day situations, such as to buy masks, gloves and disinfectant, as well as the greater need of these social groups to travel in public transports. Poor people who do not have access to health services under normal circumstances are more vulnerable in times of crisis. The manipulation of information, poor content quality and difficulties in accessing communication and information technologies affect individuals with fewer resources and enhance illiteracy, making them more likely to ignore the health warnings developed by the governments [11,29].

In several discursive records, widely disseminated in the global public space, the idea is conveyed that this is a moment of unpredictable and disturbing rupture: a "health crisis", an "international health emergency", using WHO's expressions. The notion of crisis entails a normal state and its temporary disturbance before a return to normality, i.e., the emphasis is placed on a linear view of the crisis [30]. A crisis has several properties: the loss of meaning, the de-sectorisation and its complex, urgent and dynamic nature. These three dimensions are combined. A crisis is a test of an existing order and, in particular, of cognitive categories and actions, but also of the limits and hierarchical structures that organise it, which can lead to organisational collapse [31–35]. Another challenge underlying the widespread use of the "health crisis" notion lies in the appreciation of the noun "health". By qualifying the event in this way, priority is ascribed to the health framework over others. However, one of the effects of these "crises" is to go beyond the usual organisation by sectors, producing a "de-sectorisation" [32]. This forces people to think outside their sole area of competence and forces coordination. The health, social, ecological, economic, financial and political dimensions of a "pandemic crisis" are interdependent.

The pandemic, as a global threat, is therefore imbued with a constant state of crisis, in which the crisis becomes permanent and the cause that explains everything [36]. This global threat has an epidemiological and medical dimension, but also a political and governance one. The clinical and epidemiological approaches imbricate to a political component (power, violence, constraint) and a governance component (state structure, governments' behaviour). The State's mastery of threats depends on its capacity to create, develop and manage complex and specialised organisations (care and health system, agencies and expert committees, among others), its capacity to ensure the continuity of its functioning and the mobilisation of its resources, as well as its power to control the use of coercion in response to global threats and dangers [37].

What added to the crisis was the decision of the authorities in many countries to resort to general population containment for a long period [31]. There are no studies that allow political actors, when taking this type of decision, to anticipate its consequences for the physical and mental health of populations, the relationships within households, the care of dependent, isolated or precarious individuals, the economy, work, the life of organisations or education. Governments only have hospitalisation and mortality rates as indicators to guide their actions. To this loss of meaning can be added the confusion in crisis management between the different authorities at the top of the state, as the borders are unclear and give rise to jurisdictional struggles [31].

This economic, social and health crisis has serious and profound implications that raise questions to which the contribution of Social Sciences does not seem to be sufficiently mobilised by policy-makers (research in SCILIT, virtual social networks and other databases) [38]. Social Sciences can collaborate more intensely in better knowing and managing this epidemic [20,22,39–41].

Besides studying society, Social Sciences are also part of it. They develop a permanent self-reflexivity as a social practice and system of representations, stage of conflicts of interest and power games, on the practices of Social Sciences as a scientific and professional activity socially conditioned, socially produced and always with social consequences [42,43]. This reflexive capacity establishes the domain of the symbolic and becomes effective to the extent that it provides cognitive and representative elements of adaptation to reality. The knowledge and representations that individuals, groups and societies have and use—generally referred to as "common sense"—are, in modern societies, increasingly shaped by Social Sciences. This social appropriation of knowledge developed by Social Sciences has a return effect by causing a permanent rethinking of problems and conceptual elaborations [44]. The potentials of this reflexivity frame the reflection on several domains of the "societies of individuals" [45] affected by this permanent crisis.

The objective of this concept paper focuses on the relevance of the analytical potential of Social Sciences for the understanding the multiple implications and challenges posed by the COVID-19 contagion–pandemic dyad. This crisis has profound implications and raises issues for which the contribution of Social Sciences does not seem to be sufficiently mobilised. In the analysis, the contribution of Social Sciences is paramount, in terms of their knowledge and skills, to the knowledge of these problematic realities and to act in an informed way on these crises. Social Sciences are a scientific project focused on interdisciplinarity, theoretical and methodological plurality. This discussion is developed from the systems of relationships between social phenomena in the coordinates of time and place, and in the socio-historical contexts in which they are integrated. A pandemic is a complex phenomenon as it is always a point of articulation between natural and social determinations. The discourse space on the COVID-19 pandemic can be understood as the expression of a coalition of discourses. The circumstantial coalitions of interests, which shape the different discursive records and actions produced by different agents of different social spaces enable the acknowledgement and legitimation of this pandemic threat and danger, and the promotion of its public management. It is, therefore, important to promote an interdisciplinary scientific project characterised by the interdependence between the epidemiological, medical and biological knowledge and the knowledge produced by the Social and Human Sciences to better understand an economic, social and health crisis of such a huge scale and to shape the medical and political management of this and future epidemics and pandemics. To attain this goal, the article is divided into the following sections: Methods; COVID-19: the contagion–pandemic dyad; multiple implications of the COVID-19 pandemic; and Conclusion.

2. Methodology

The methodology used in this article is qualitative, as the article seeks to understand the multiplicity of facets and dimensions that characterise the plural discursive space of social sciences about the COVID-19 contagion-pandemic dyad. This discursive space may be envisaged as the expression of a coalition of discourses, that is, the interaction of several discourses, combined in re-interpretative modalities of certain realities and social phenomena.

The document analysis was the favoured technique in this research, developed from different types of documentary sources. In a documentary study, documents can be understood as "means of communication", created with a purpose to attain a goal. They can be seen as a way to contextualise information, and are analysed as communicative devices methodologically developed in the production of versions about events [46]. Considering that documents are instruments and communication supports that express objectified forms of experiences and knowledge related to a specific sector of human practices, the whole document is likely to be contextualised in a specific social and cultural framework. In this context, of a complex nature, documents are inserted into the framework of social relationships and play a given role in the game of social relationships, ascribing value to acts or shaping relationships [47].

The type of the selected documentary sources encompasses articles and interviews, produced by several scientists in the fields of social sciences, epidemiology and public health. When expressing different stances and interests in the scientific, political and societal fields, these authors were important in the understanding of the construction of the discursive space on the social configuration of the COVID-19 pandemic. These documents create and display cognition policies: the language is used to build an "official" reality—often on classifications that produce the promotion or marginalisation of ideas—and the contemporary concepts, which define the situations and shape the readers' preferences, perceptions and cognitions. These documents may be viewed as places of symbolic struggles in terms of the perception of the social world. These symbolic struggles may have two different forms. The objective form shapes the possibility of acting through individual or collective representation actions, which aim to show and promote specific realities. The subjective form models the possibility of acting, seeking to change the categories of perception and assessment of the social world, the cognitive and evaluative structures: the categories of perception and the systems of classification, that is, the words and the names that build the social reality as much as they express it [48,49].

3. COVID-19: The Contagion–Pandemic Dyad

At present, COVID-19 is seen as a pandemic, a new "plague", referring to the matrix metaphor of the plague expressed in the coalescence of two semantic nuclei: contagion/death [50]. The term epidemic, which refers to a disease that suddenly affects a large number of people and can cause a high number of deaths, has a very close meaning to the Latin word *pestis*. This term emerges in the 12th century, in Latin texts, as an erudite form of designation of the social phenomenon that until then was called *pestis* [50]. In various discursive records of a religious, poetic and medical nature, the term *pestis* and its derivations designate a terrible evil, a great misfortune, or even a polluted, contaminated, unhealthy, insane and often deadly condition. It represents a contagious disease but, above all, the great evil. The domains and conceptual uses of contagion, its modalities and its effects, in its various historical contexts and meanings, express the interdependence between the positions and the different points of view of the various actors involved in multiple scientific, moral, social and political challenges.

The contagion–epidemic dyad was underpinned by the perception of great danger, amplified in contemporary societies by the extreme speed of information circulation of people and goods. The belief in the contagion of epidemic diseases and events perceived as dangerous can be identified with the symbolic dimension of impurity, which expresses the dangers to the social order and the purity of a group, emerging as a guarantee of the avoidance of any physical and social approach to the sick and other individuals perceived as dangerous in a given community. This articulation between impurity, purification and interdiction of contact by believing in contagion shapes the symbolic management of internal and external dangers that threaten the various societies in different time-spaces. This dyad is related to experiences of fear, exasperation, denial and rejection of the other, recalling the old attitudes of populations threatened by the plague, such as those of escape, purification and isolation [51]. The perceptions and representations of contagion convey negative or derogatory cognitive sensations and evaluations suggesting the malignity of the other, the other as a source of threat and danger, leading to generic fear in which contact is perceived as the possibility of aggression and contamination.

A significant number of epidemics are characterised by a mode of inter-human transmission. For such an epidemic to occur, social interactions between individuals and groups of individuals must occur. The more numerous, regular and rapid these interactions, the higher the risk that an epidemic will break out and spread on a large scale. Conversely, a pathogen will find it more difficult to transmit in a group of individuals whose interactions are weak, or between groups that have few relationships with each other.

As mentioned, in early December 2019, a new coronavirus causing potentially severe pneumonia emerged in the Asian region. Although the first cases of infection were identified in fish-market-goers, new infections were quickly detected in individuals that did not visit this market. The reservoir of this coronavirus is probably an animal. SARS-CoV-2 is genetically very similar to other animals circulating in China in the natural populations of horseshoe bats of the species *Rhinolophus affinis* [52–54]. This virus may have resulted from a natural genetic combination of two coronaviruses with two different animal hosts [53]. Some studies suggest that the pangolin, a small mammal consumed in Southern China, could be involved as an intermediate host between bats and humans. It quickly became clear that this virus could be transmitted between infected humans. Transmission between humans has been established and experts estimate that, in the absence of control and prevention measures, each patient could infect between two and three individuals [54].

The mode of transmission appears to be carried out predominantly by droplets of respiratory secretions during prolonged close contact or by direct contact when the infected person coughs, sneezes or talks, which can be inhaled or landed in the mouth, nose or eyes of people who are close by. A growing body of epidemiological data currently available on cases around the world shows that the vast majority of cases have been linked to transmission from person to person during unprotected close contact with a person with symptoms compatible with COVID-19 [52]. Coronaviruses may also remain viable (i.e., maintain their infectivity) for a few days in the environment and this period depends, for example, on environmental temperature or exposure to ultraviolet radiation. Thus, the transmission of these viruses to humans may occur when hands, contaminated by contact with, for example, surfaces where these viruses may have been deposited, are brought into the eyes, nose or mouth. Regarding specifically COVID-19, WHO estimates that the incubation period of the virus varies between one and 14 days, with an average time of five days. During this relatively long incubation period, individuals may be asymptomatic carriers of the virus and, hence, contagious agents.

When a person is in contact with a pathogen, such as a virus or a bacterium, his or her body establishes a defence process: the body's immune system produces antibodies, specific molecules of the pathogen, which bind to it and destroy it. These antibodies continue to be produced and circulate in the bloodstream after healing: it is the principle of "individual immunity", which allows the body to protect itself from a second infection by the same pathogen. "Group immunity" derives from this principle; if a high proportion of people in a given population have been infected with a virus, the individuals will become immune to the pathogen, breaking its transmission chain. The proportion of the people needed to establish group immunity in a given population varies according to the pathogen involved and the degree of contagiousness of the disease. This degree of contagion is measured by the term R0, which indicates how many new infections are caused by each case. For COVID-19, the R0 is estimated to be 3.28 in the UK, although this figure is likely to change because studies are still ongoing. This means that, to have group immunity, about 70% of the population from the UK would have to be immune to COVID-19 [55].

Social interactions on a global scale shape pandemics [56]. Rather than progressively globalising from one scale to another—from local to global to national and regional—global epidemics emerge directly on a global scale and coexist with other epidemics on lower scales [57]. This argument favours three elements: (i) epidemics spread thanks to co-presence and mobility; (ii) epidemics spread according to a mainly topographic logic; and (iii) the spread of world epidemics is mainly topological [56].

Societies have three means of optimising their interactions: (i) co-presence, i.e., the gathering of several individuals in the same place; (ii) mobility, which corresponds to the movement from one place to another; and (iii) telecommunications, which enable individuals to come into contact with one another without moving. Co-presence and mobility play a critical role in the spread of epidemics. First of all, the large concentrations of individuals that cities are allow a pathogen to have access to a large number of potential hosts and, thus, be able to evolve more rapidly than the progressive immunisation of populations. Transport enables pathogens to spread more quickly and further but also to reach non-immunised populations.

The spread of pandemics is mainly topological. The spread of HIV/AIDS (Human Immunodeficiency Virus/Acquired Immunodeficiency Syndrome) and SARS has developed mainly according to topological logics, in which the notion of proximity is defined in terms of connectivity between different points of a network [56]. The rules of diffusion entail that a site strongly linked to the point of origin of the epidemic will, usually, be affected before a weakly linked site, regardless of the distance in kilometres; sites affected at the same time from the same starting point of the epidemic have a similar degree of connection (they may, then, be thousands of kilometres away from each other). The topographic and topological diffusion or network logics are interdependent. Thus, it is possible to distinguish three scales of diffusion for the same epidemic: national, regional and global [56]. These three scales are present in the process of dissemination of the COVID-19 pandemic: one worldwide, which spread from the city of Wuhan within the Megalopolitan World Archipelago [57]; one regional, which mainly affects Southern Europe, the United Kingdom, the United States of America, Mexico and Brazil; and one national, whose features may vary in each affected country.

The literal or metaphorical sense of the notion of contagion can be mobilised to qualify the distribution of a medical and/or social phenomenon, characterised by its modalities of expansion, transmission and distribution, by an exponential propagation speed, sometimes associated with an unexpected origin and an unpredictable end. The notion of contagion often expresses fear, vulnerability, ignorance and powerlessness but also implies the mobilisation of warning procedures, as well as prevention and preparation devices, to deal with these disruptive situations. The discourses on contagion shape the normative justifications, developed by the institutions to formulate and implement coercive measures, namely isolation, quarantine, social distancing and exalting the saving qualities of vaccines. Naming contagion allows for the naturalisation of a threat, which an uncontrolled proliferation of an illness or disease poses in a particular community, justifying the adoption of measures to manage this threat [3].

4. Multiple Implications of the COVID-19 Pandemic Virus

The COVID-19 pandemic can be regarded as the first pandemic of the globalisation era since it articulates certain features: global scale, extremely fast transmission speed, cross effects of global inter-territorial interdependences, the interdependence of nations in the management of their respective epidemics, and growing complexity in the spatial organisation of economic globalisation [58]. Pandemics may be viewed as one of the main risks of modernisation. The consequences of scientific and industrial development are a set of risks that cannot be contained in space or time, for example, in addition to ecological risks, environmental degradation (causing viruses spill-over between different species), the exponential development of the transport network (commercial aviation has been instrumental in the rapid spread of pathogens). Furthermore, there is a change in patterns and lifestyles and an exponential increase in the consumption of animal protein [59], increasing and massive precariousness of living conditions, with individualisation of social inequality and uncertainty about employment conditions, leading to widespread exposure to risks [60]. While increasing inter-population connectivity, globalisation increases in scale and intensity of action and impact.

Due to the impact of global risks, countries, even in the West, are more authoritarian but inefficient when it comes to dealing with different global threats and hazards, becoming "strong failed states" [60] (p. 79). This is also linked to the issue of insurance and the degree of risk control. Private insurance is coming to an end, and the state is always the ultimate guarantor of the value of goods and people's lives. The state thus takes on a central role at a time when most risks are global [60,61].

The COVID-19 pandemic is not an unprecedented phenomenon. It is part of a process of emergence and re-emergence of infectious diseases, which has presented many severe episodes in recent decades. Some examples are Influenza H5N1 in 1997 and 2005, SARS in 2003, H1N1 in 2009, MERS in 2012, and the Ebola epidemic since 2014. These epidemic phenomena have been characterised by significant health risks for populations and health professionals, given the high virulence and high degree of exposure of these professionals, in particular the SARS epidemic. This epidemic has given rise to a twofold surprise: on the one hand, the emergence of a new pathology with an unprecedented rate of spread and, on the other hand, the refutation of a dogma rooted throughout the 20th century, according to which the control of infectious phenomena in the developed world would protect it from health disasters, while the corollary of the significant increase in risk would only affect developing countries [62].

At the beginning of the 21st century, WHO encouraged countries to prepare for a severe pandemic and its occurrence was deemed very likely from 2003 onwards. It called on all countries to establish national pandemic planning committees [63]. A kind of revolution in health policy would be underway, adding a culture of preparedness to a culture of prevention (disease prevention). Preparedness covers a set of notions, strategies and activities, and can be systematically defined around three axes: prevention of major epidemics through epidemiological surveillance and early detection of epidemic outbreaks; establishment of rapid responses, producing and sharing data, knowledge and technology; and improving the global governance of the response to epidemics [64–66]. Several procedures have been adopted for the preparation of a health system for large-scale health threats. In some countries, public bodies have decided to create and manage a national health reserve: medicines (antivirals, vaccines) and materials (various types of masks, among others). For example, in France, it was decided to finance a national mask industry not to depend on uncertain imports, particularly from China [63]. These plans for intervention and the organisation of countries to reorganise societies and maintain the essential functions of the economy were developed integrating the main WHO recommendation: the need to mobilise society as a whole in this process of preparation [63].

However, the H1N1 epidemic in 2008–2009 had a paradoxical effect. Heralded as terrible, the pandemic was considered much less serious, despite causing over 200,000 deaths worldwide. The failures of global and local governance to prevent and regulate permanent financial crises, environmental risks, the crisis in international institutions and the inherent increase in geopolitical risks have led to demobilisation and budget cuts, reducing these preparedness devices to almost nothing, an expression of general political indifference.

This volatile situation has shaped thinking and managing health threats and the subsequent response of individual countries to the COVID-19 pandemic. Not only has there been no upstream preparedness, but it is precisely the idea of crisis preparedness that seems to be totally absent from government actions. The measures that are too limited or too late may be the subject of criticism but, more fundamentally, it is the consequences of this or that measure taken after the declaration of a state of emergency that had not been foreseen, precisely because of lack of preparedness [63]. Faced with this "anomaly", one could expect a scientific and political consensus between various actors for its political and medical management, the definition of urgent and collective actions and the demand for a common defence, surveillance and prevention logic by the States involved in a pandemic threat. In the face of this threat, the variations in States' preventive or prophylactic strategies are pure products of politics, of the effects of the nature of the regimes in force, as much as—if not more than—the circumstances surrounding the very course of the epidemic [37]. In the areas of national sovereignty, in particular, concerning the policies of confining populations and closing borders, there is weak coordination between the various countries.

The management of public health crises, of emergencies, has gained a political dimension, rather than a scientific or technical one because health problems are extremely sensitive to the public and include major political risks. In this political management, the relationships between political decision-makers, experts and economic agents are critical [67], knowing that the uncertainty of the context of science intersects with the interpretative uncertainty of each of the actors involved [68].

The impacts of the COVID-19 pandemic are multiple: social, economic, health, political, educational, labour and employment, ethnic-racial, on freedom and on citizenship rights. In addition to these multiple impacts, they are interlinked in complex, sometimes paradoxical ways and produce inequalities, social injustices and different types of discrimination [29]. Individuals, groups and countries are not all equal in their vulnerability to disease. The pandemic crisis has led to the worsening of social inequalities in the global context, between countries and within countries; the most disadvantaged and vulnerable ones are increasingly disadvantaged and vulnerable. In several social categories, there are situations of increased inequality, and others are affected by new situations of inequality: (i) older people, who belong to one of the high-risk groups of infection with the coronavirus, are less prone to use distance communication tools and are more likely to be socially isolated; (ii) young people, who directly suffer the effects of disrupted education in these troubled times and, hence, may have more difficulty of find a good job upon conclusion of their studies; (iii) women, who are more susceptible to become children or elderly carers and may have higher risk of being victims of domestic violence; (iv) individuals from East Asia, who may be at higher risk of discrimination and even persecution due to the fact that COVID-19 is associated with East Asian countries, notably China; (v) individuals with reduced communication competences and literacy, who may not receive or understand official communications on the pandemic; (vi); homeless people, who may find it hard to self-isolate or seek for support; and (vii) workers on precarious contracts or self-employed, who have higher risk of reduced to no income [24].

There is a fundamental difference in the vulnerability to the pandemic between the countries that have social states based on the principle of universality in access to health care and social protection and those where such states do not exist. The management of COVID-19 by health systems is shaped by several relevant elements, namely the degree of contagion, the degree of exposure to the market of people and countries in access to health care, and community-based management of the disease. A significant resource constraint, associated with a scenario of continued downgrading and submission of other relevant health activities, has characterised this pandemic crisis management process, leading to the neglect of all other diseases, including chronic, oncologic, cardiovascular and mental diseases, which could culminate in a significant increase in morbidity and mortality rates [59].

Another significant impact of this crisis is on the operating conditions of democracy. The response to the pandemic raises the restriction of freedom of movement, containment of activities, the strengthening of the countries' executive powers. In some countries, the strategy of governments has been to annul democratic order, suspend the functioning of parliaments and attempt to govern by decree, and trivialise exceptional symbolic resources—e.g., the use of the armed forces to control public order [69]. The world is witnessing the consolidation of some features of technocratic authoritarianism, which is very present in China, namely the attraction of the electronic means of citizens' permanent surveillance—surveillance of contact networks and surveillance of infected people, among others.

The first assessments of the economic impacts of the pandemic crisis point toward a drop in global trade of goods by 2020 of more than 35%, a drop in tourism revenue flows of more than $500 billion, and losses in commercial aviation of close to $70 billion. The fall in economic activity could range from −3% for the world economy, −6% for the United States economy and −7.5% for the euro area [70]. IMF (International Monetary Fund) estimates a drop of 3% in global GDP (Gross Domestic Product) in 2020, with will be much sharper in advanced economies (−6.1%) than in emerging or developing economies (−1.0%) [71]. In 2020, around 90% of the countries in the world will face a recession. For the Euro Zone, it is estimated that the unemployment rate will increase from 7.6% to 10.4% in 2020, representing the loss of around 4.5 million jobs. In the United States, the unemployment rate, according to these estimates, almost triples, reaching 10.4% in 2020. In China, the unemployment rate is expected to increase by only 0.7% in 2020 [70–72].

Almost all countries have implemented strict movement controls to cope with the COVID-19 pandemic. The aim is to prevent transmission by reducing contact between people and to slow down the spread of infection by reducing the risk of public health services, which will be overloaded. However, these measures may also prolong the pandemic and the restrictions adopted to mitigate it [24].

5. Social Sciences and COVID-19

The contributions of the perspectives developed by the various Social Sciences on the disease, contagion and epidemics remain in a relatively secondary position in the public arena in this process of pandemic crisis. On the contrary, epidemics seem to be a prerogative of the Hard Sciences, where biological and epidemiological categories are repeatedly mobilised to explain the epidemic phenomenon and often also to analyse historical and social phenomena. Several authors, relying on certain epidemiological methods to analyse the factors contributing to the general history of human populations, maintain that the movement of populations, their evolution and mutations would be causally dependent on epidemics, the history of microbes and viruses. This approach would allow a posteriori modelling of the determinants of population evolution [73,74].

The analysis of social determinants of health involves the deconstruction of biomedical assumptions in the analysis of social determinants of health: the epidemiological risk factors. Risk does not only relate to the knowledge of a given pathology, its transmission and evolution. Risk is a central category in the management of public problems, enabling responses to be shaped to the social representations of risk and to the uncertainties and possible dangers which individuals are exposed to, by the classifications and identifications it operates. In pandemic threats, individuals face competing risks. Some people are willing to expose themselves to the risk of being infected by the coronavirus because they do not comply with the containment measures, inasmuch that the immediate competing risks weigh more heavily: they run the risk of losing their jobs, their income; they fear loneliness [75].

Disease and health, regardless of the biological and physiological dimensions that make up their medical and clinical realities, can be regarded as symbolic maps that reveal the political, social and cultural structures that ascribe them meaning, involved in multiple regulatory regimes. The establishment of moral boundaries supports the process of categorising and labelling diseases, based on the functioning of the binary model, which, for example, opposes decent diseases to indecent diseases; shameful diseases to morally acceptable diseases; pure diseases to impure diseases; polluted diseases to impure diseases; clean diseases to dirty diseases [76]. Epidemics and contagion reinforce prejudices and enhance stigmatisation processes: against the Jews (the 1347–1350 plague), against the poor in the Renaissance (the plague, typhus), against Irish immigrants in the 19th century (cholera), against the poor in the 19th and 20th centuries (tuberculosis), against the four Hs (homosexuals, Haitians, haemophiliacs and heroin-maniacs), in the 1980s (HIV/AIDS), against Africans (Ebola), against Chinese and Asians (bird flu, SARS and the coronavirus), against the "stranger", the "foreigner": the "other" ...

The pandemic configuration is shaped by interdependent relationships between social structures and individuals. Social behaviours, structures and cognitive assessments are influenced by the unfolding of pandemics, as is the case of COVID-19 [8], with profound implications for the organisation of a society [20]. Table 1 offers some examples of this social dimension.

The measures adopted by many countries and regions to establish quarantine, social isolation and community containment have profound implications beyond the purely medical and public health aspects in controlling this disease. On the one hand, they raise questions about individual freedom and human rights [14,23,78–83]. On the other hand, social interactions and social institutions are central elements in the control of this disease [7,14,40], which brings out the importance of the Social Sciences' perspectives. For example, concerning social interactions, in Southern European countries intergenerational interactions are more frequent than in other parts of Europe since social norms about providing support to family members and maintaining interpersonal family interactions are stronger than in Central and Northern European countries, which are less family-oriented [40].

Table 1. Impacts of the pandemic dynamics.

1	Communities at various scales (i.e., towns, cities, nations, regions) with higher levels of economic wealth, more equitably distributed temporally and spatially, will generally tend to be more resilient to existential and other threats.
2	Where communities are comparably affluent and on an increasing trajectory of affluence, those with more equitably distributed wealth and higher floors of minimum wealth are probably more likely to be more resilient.
3	From a community development (i.e., municipal, state, regional) and public infrastructure perspectives, it appears socially-equitable economic development leads to more efficient outcomes, with respect to system resilience, than does inequitable economic development.
4	Both socially-equitable and socially-inequitable development scale up, and hence, communities can learn and implement best and effective practices for socially-equitable development as a resilience strategy for infrastructure, economic and community development at higher scales.
5	The fact that climate, pandemic and other threats are no respecter of persons elevates the importance of socially-equitable development as a resilience strategy.
6	Infrastructure policymaking, planning, design and construction, maintenance and renewal approaches, if they do not formally incorporate equity considerations, will lead to the development of less resilient infrastructure and communities, with avoidably higher opportunity costs.
7	Social equity is intimately intertwined with economic advancement. Socially-equitable economic development can foster not only socially-equitable advancement that creates better conditions for all but also support the development of system resilience-economic, social and infrastructure.

Source: Amekudzi-Kennedy, Labi, Woodall, Marsden and Grubert [77].

Thus, according to social scientific insights, several responses to COVID-19 should be encouraged for more efficiently dealing with the pandemic. Among them, van Bavel et al. [41] (p. 2) identify and highlight the following:

- A shared sense of identity or purpose can be encouraged by addressing the public in collective terms and by urging "us" to act for the common good.
- Identifying sources (e.g., religious or community leaders) that are credible to different audiences to share public health messages can be effective.
- Leaders and the media might try to promote cooperative behavior by emphasizing that cooperating is the right thing to do and that other people are already cooperating.
- Norms of prosocial behaviour are more effective when coupled with the expectation of social approval and modeled by ingroup members who are central in social networks.
- Leaders and members of the media should highlight bipartisan support for COVID-related measures, when they exist, as such endorsements in other contexts have reduced polarization and led to less biased reasoning.
- There is a need for more targeted public health information within marginalized communities, and for partnerships between public health authorities and trusted organizations that are internal to these communities.
- Messages that (1) emphasize benefits to the recipient, (2) focus on protecting others, (3) align with the recipient's moral values, (4) appeal to social consensus or scientific norms, and/or (5) highlight the prospect of social group approval tend to be persuasive.
- Given the importance of slowing infections, it may be helpful to make people aware that they benefit from others' access to preventative measures.

- Preparing people for misinformation and ensuring they have accurate information and counterarguments against false information before they encounter conspiracy theories, fake news, or other forms of misinformation, can help 'inoculate' them against false information.
- Use of the term "social distancing" might imply that one needs to cut off meaningful interactions. A preferable term is "physical distancing," because it allows for the fact that social connection is possible even when people are physically separated.

The processes of knowledge denominate and qualify social reality at the very moment of analysis, and disciplines establish dimensions of that reality when analysing it. The plurality of Social Sciences results, on the one hand, from the real complexity of human action, which demands complex analytical levels and elaborations, and, on the other hand, from the approximate and parcelled nature of scientific knowledge, which denies the possibility of unitary knowledge of the real [84]. Seeking to know social reality entails the mediation of categorical frameworks, logical operators of classification, ordering—as complex processes influenced by our needs, experiences and interests—that provide information about this reality and ways of making it intelligible, but confusing with it [84].

Bearing in mind that each Social Science has a specific perspective on social reality, the pandemic crisis, as a complex and multidimensional phenomenon, can be understood from different perspectives. This discussion would be developed from the systems of relationships between social phenomena in the coordinates of time and place, and in the socio-historical contexts where they are integrated. The conceptions of common sense and ideologies about these processes would also be the object of analysis [84]. Social Sciences are crossed by large, distinct, partially overlapping and partially contradictory theoretical configurations that guide empirical research from different paradigms that preserve zones of specificity and, also for this reason, can lead to results and rival explanations of the reality they analyse [44].

The investment of political actors in the collective management of epidemic diseases has enabled the emergence of the autonomous public health area. Politicians can be regarded as power relationships mobilised in the public space for the control of decisions and actions that have as their object assets deemed collective [85]. It may be questioned whether the political management of this pandemic crisis expresses a reconfiguration of the social space of public health, in which surveillance, monitoring, evaluation and risk management would have a place of growing importance, supported by technologies of visualisation, tracking, analysis and a "molecular vision of life", which would be associated with a division of labour, differentiating and fragmenting the "objects" of medicine. Health practices would be increasingly shaped by the demands of a new political health economy, subject to the imperatives of large groups and companies, insurance companies and, in general, the world of bio-capital and would reach the public health systems themselves [86](p. 150).

In the response to the pandemic, inequalities have not been mitigated but rather reinforced. This situation leads to the equation of social inequalities in the near future from three axes: inequalities and the social state; inequalities and digital transformation; and inequalities and science [29]. Regarding the first axis, the institutions of the social state, mainly the National Health Service, the public social protection system, retirement and old-age pensions, unemployment and sickness benefits, are pivotal in containing and possibly reducing social inequalities. Two priorities became evident with the pandemic crisis: the strengthening of public systems; and the accentuation of their universal nature [29]. The policy response to the dynamics of employment and unemployment will determine the evolution of inequalities in the near future [69]. In the second axis, the ambivalence of the digital transformation stands out. While digital devices have great potential in several areas, they can generate major threats to the ways of human existence in society. The attraction of a surveillance society for the sake of effectiveness in public action, particularly in crisis situations, with the use of permanent electronic surveillance of individuals, can configure technocratic authoritarianism. There is a concentration of data in a very limited number of digital companies in some states at the centre of the world-system. When the interests of the state and private interests in the use and control of information and communication technologies—as the potential technological basis of totalitarianism—come together, the risks to democracy become very high [69]. Algorithms are imbued with prejudices and discriminatory criteria. A social stratification managed by digital means, directed by political-administrative hierarchies, is consolidated, with widespread social unequalled effects [29]. The third axis equates the utility, credibility and trust ascribed by the various "societies of individuals" to science and scientists in the production of knowledge and the definition of strategies to find solutions to the disease. The growing social visibility of science and its public communication may enhance the affirmation, simultaneously, of "scientific culture" as a citizen's right and of a more reflexive conception of science, which has the purpose of critically analysing what it does, to assess the conditions in which it does it, as well as the effects of its activity [42]. The generalisation of the use of Information and Communication Technologies in scientific work shapes the ways of doing science and the emergence of new and wider modalities of science communication. One of the main changes is the breaking of communication linearity, replaced by an interactive network model, enabling higher levels of collaboration, internationalisation, transparency and impact of scientific work [87–89].

The observation that social interactions between individuals on global scale shape pandemics, a "global crisis", a "global threat", encourages reflection on the global nature of epidemics and, more broadly, health. A minimalist version assimilates "global" to "universal", i.e., it reduces the phenomenon to a simple spatial extension; in the infectious field, it is the passage from the epidemic to the pandemic. Yet, if this is the favoured meaning, it would make little sense to replace "international health" with "global health", as WHO did at the beginning of the 2000s, when its legitimacy was diminishing in the face of the rise of large private players [90]. Globalisation is not a homogeneous process: disease, medical response and social perception evolve in a differentiated way. Globalisation has less to do with extension than with interconnection; it has to do with networks of individuals, organisations, ideas and effects. In these networks, the mobility of individuals spreads the disease, the interactions between public and private institutions determine the response, and the relationships between the media and their audiences are part of the formation of perception, amplifying it or, conversely, reducing it, as is the case of some diseases, which, despite their high morbidity and mortality rates, no longer instigate panic and marked fear [90].

The social history of epidemics and the cultural studies on epidemiology, contagion and infectious diseases have highlighted a fundamental challenge facing the various Social Sciences: how to live together [91], what bonds us and what separates us. One of the most immediate effects in any epidemic outbreak is the material and symbolic exacerbation of social differentiation, the multiplication of the dividing lines between "us" and "the others": between the healthy and the sick; between those who are well and those who are not; between those who have "previous pathologies" or are part of "risk groups" and those who promote "healthy" lifestyles; between those who have resources and support and those who do not; between "those who are inside" and "those who are outside". These differences slide very easily into a social discourse towards a distinction between the "innocent" and the "guilty" [76,91,92].

Isolation has not only broken the regularity of social ties and connections in co-presence and physical proximity but is also generating a set of disruptive processes emerging from prolonged confinement [93]. In the long run, social isolation is associated with an increase in mortality of almost one third [94]. Prolonged periods of social isolation may have similar effects [24,95]. Individuals who are socio-economic disadvantaged or with poor physical or mental health are at higher risk [24]. There is no decent society that is based on the disconnection and interruption of social relationships, as well as on the drastic deepening of socio-economic vulnerabilities. This pandemic situation stands for an abnormality that cannot (should not) become the new normal. If this happens, it is the very notion of society that is at stake [93].

One of the most pressing challenges facing Social Sciences and Humanities in this ongoing crisis is to understand the extent to which the morality of our actions is determined by the value of their consequences. Dilemmas such as leaving a significant part of the population of a given country to contract the disease to develop collective immunity to prevent future epidemic outbreaks, or deciding who deserves to die or live in a context of a relative scarcity of resources (beds, masks and ventilators, among others) are moral problems faced by the agents responsible for the medical and political management of the disease [96].

These situations, among socially relevant others, relate to the relevance of the fundamental contribution of Social Sciences, their knowledge and skills to the knowledge of these problem realities and to act in an informed manner, respecting democratic freedoms, human rights and the sense of social justice, and in containing and solving the problems posed in contemporary societies by this and future economic, social and health crises [29]. Table 2 presents some of the Social Science fields and their role in the management of the COVID-19 pandemic.

Table 2. Social Sciences in the management of the COVID-19 pandemic.

Psychology	Mental health consequences of social isolation; analysis of psychological resilience
Social Psychology	Social representations on the pandemic; attitudes towards risk; health attitudes and behaviours; assessment and intervention in adaptation to disease and health promotion; violence and social confinement; social ties and new forms of solidarity
Political Science	International governance and multilateralism; governance and public policy; political institutions, political attitudes and behaviour; public opinion and political communication; pandemics and nationalism
Economy	Dynamics of capitalism and pandemic; economy and health; structure of employment and unemployment; macroeconomic vulnerabilities; Lay-off; teleworking
Geography	Genesis and pandemic spread of SARS-Cov-2; metropolises, city networks and pandemic spread; migrations and pandemics
Demography	Demographic structure and prevalence of the disease; demographic vulnerability and COVID-19
Communication	Digital transformation, ICT and health surveillance; digital literacy; telemedicine and teleconsultation
International Relations	Articulation and/or competition between countries in fighting the pandemic
History	History of epidemics, pandemics and contagion; history and collective management of the disease and contagion
Anthropology	Cultural analysis of risk and contagion; processes of stigmatization of the "other"; cultures of fear; an ethnography of zoonoses
Sociology	Social order and crises; social inequalities in health; social construction of epidemic and contagious diseases; social space of health and collective management of disease and contagion; public health policies; social distance and confinement; science and scientific controversies

Source: Authors' own production.

6. Conclusions

A pandemic is a complex phenomenon as it is always a point of articulation between natural and social determinations. Its analysis is transversal: it is important to capture the points of intersection of the determinations and analyse their consequences. In this process, Social Sciences assume themselves as a scientific project centred on interdisciplinarity that favours a relationship of interdependence and complementarity between Social Sciences, in the theoretical and methodological plurality that seeks to articulate macro-social dynamics with local processes, allowing the connection of subjective meanings with practices, and that focuses on the articulations between systems and actors, between structures and practices, between the reality of social conditions of existence and the social construction of reality [97–100]. This enables the demystification of common knowledge shared in society, but which in reality is not correct [101,102]; and fosters sociological imagination [103].

The discourse space on the COVID-19 contagion–pandemic dyad can be understood as the expression of a coalition of discourses, i.e., the interaction of various discourses, combined in re-interpretative modalities of certain realities and social phenomena. The circumstantial coalitions of interests, which shape the different discursive records and actions produced by different agents of different social spaces (political, medical, scientific, economic and religious)—enable the acknowledgement and legitimation of this pandemic threat and danger— and the promotion of its public management. They add to the definition and implementation of health policies, as well as to the promotion and development of institutional areas of a preventive nature, namely vaccination, health cordons, quarantine, isolation, social distancing, and intervening in the configuration of the collective management of disease and health.

In the case of pandemics, authors such as von Braun, Zamagni and Sorondo [104] argue that understanding the meanings of various social actions is critical in their management [20,105], with the development of specific measures for different social groups [39,40,77] and the possibility to counter the logics underlying the stigmatisation of the "other" [106]. In the COVID-19 pandemic, they imbricate: a new virus (SARS-CoV-2, with its own genetic characteristics); a human actor who is both the carrier of the pathogen and the diagnosed (or not) patient who brings out illness and sickness; an environment where nature (animal reservoir), humans and society interact; the knowledge of scientific medicine that is created at the same time as the virus replicates and the disease produces its effects, and legitimises public policies—a link that gives rise to an irreducible, inextricably biological, environmental and social relationship [107,108]. It is, therefore, important to promote an interdisciplinary scientific project characterised by the interdependence between the epidemiological, medical and biological knowledge and the knowledge produced by the Social and Human Sciences to better understand an economic, social and health crisis of such a huge scale and to shape the medical and political management of this and future epidemics and pandemics.

Author Contributions: Conceptualization, C.M.F., M.J.S., J.G.M. and S.S.; methodology, C.M.F., M.J.S. and S.S.; formal analysis, C.M.F., M.J.S., J.G.M. and S.S.; investigation, C.M.F., J.G.M. and S.S.; writing—original draft preparation, C.M.F., M.J.S., J.G.M. and S.S.; writing—review and editing, M.J.S. All authors have read and agreed to the published version of the manuscript.

Funding: University of Azores, Interdisciplinary Centre of Social Sciences—CICS.UAc/CICS.NOVA.UAc, UID/SOC/04647/2020, with the financial support of FCT/MEC through national funds and when applicable co-financed by FEDER under the PT2020 Partnership Agreement.

Acknowledgments: We would like to thank to the Editor and Reviewers for their comments and suggestions.

Conflicts of Interest: The authors declare no conflict of interest.

References

1. Kirksey, S.; Helmreich, S. The emergence of multispecies ethnography. *Cult. Anthropol.* **2010**, *25*, 545–576. [CrossRef]
2. Houdart, S.; Thiery, O. (Eds.) *Human, Non-Human. Comment Repopuler les Sciences Sociales [Human, Non-Human. How to Repopulate the Social Sciences]*; La Découverte: Paris, France, 2011.
3. Coste, F.; Minard, A.; Robert, A. Contagions. Histoires de la Précarité Humaine. Tracés [Contagions. Stories of Human Precariousness. Tracks]. *Rev. Sci. Hum.* **2011**, *21*. [CrossRef]
4. Ye, Z.W.; Yuan, S.; Yuen, K.S.; Fung, S.Y.; Chan, C.P.; Jin, D.Y. Zoonotic origins of human coronaviruses. *Int. J. Biol. Sci.* **2020**, *16*, 1686–1697. [CrossRef] [PubMed]
5. Fressoz, J.-B. Le Virus du Politiquement Correct. [The Virus of Political Correctness]. EHESS. Le Monde. 2020. Available online: https://www.lemonde.fr/idees/article/2020/03/25/le-virus-dupolitiquement-correct_6034334_3232.html (accessed on 25 March 2020).
6. Mohammadi, M.; Meskini, M.; Nascimento Pinto, A.L. 2019 Novel coronavirus (COVID-19) overview. *J. Public Health* **2020**, 1–9. [CrossRef]
7. Oksanen, A.; Kaakinen, M.; Latikka, R.; Savolainen, I.; Savela, N.; Koivula, A. Regulation and trust: A social science perspective on COVID-19 mortality. *SSRN Electron. J.* **2020**. [CrossRef]

8. Reher, D.S.; Requena, M.; Santis, G.; Esteve, A.; Bacci, M.L.; Padyab, M.; Sandström, G. The COVID-19 pandemic in an aging world. *SocArXiv Pap.* **2020**. [CrossRef]
9. Bezerra, A.; Silva, C.E.M.; Soares, F.; Silva, J.A.M. Factors associated with people's behavior in social isolation during the OVID-19 pandemic. *Ciência Saúde Coletiva* **2020**, *25*, 2411–2421. [CrossRef]
10. Nelson, W.; Osman, M. Sociological and psychological insights behind predicted changes as a result of COVID-19. *PsyArXiv Pap.* **2020**. [CrossRef]
11. Ahmed, F.; Ahmed, N.; Pissarides, C.; Stiglitz, J. Why inequality could spread COVID-19. *Lancet Public Health* **2020**, *5*. [CrossRef]
12. Musinguzi, G.; Asamoah, B.O. The science of social distancing and total lock down: Does it work? Whom does it benefit? *Electron. J. Gen. Med.* **2020**, *17*. [CrossRef]
13. Erol, S.; Ordoñez, G. Social and economic distancing. *SSRN Electron. J.* **2020**, 1–36. [CrossRef]
14. Sailer, M.; Stadler, M.; Botes, E.; Fischer, F.; Greiff, S. Science knowledge and trust in medicine affect individuals' behavior in pandemic crises. *SocArXiv Pap.* **2020**. [CrossRef]
15. Lunn, P.D.; Timmons, S.; Barjaková, M.; Belton, C.A.; Julienne, H.; Lavin, C. Motivating social distancing during the COVID-19 pandemic: An online experiment. *PsyArXiv Pap.* **2020**. [CrossRef]
16. Kupferschmidt, K. The lockdowns worked—But what comes next? *Science* **2020**, *368*, 218–219. [CrossRef]
17. Yu, J. Open Access Institutional and News Media Tweet Dataset For COVID-19 Social Science Research. *SocArXiv Pap.* **2020**. Available online: https://arxiv:abs/2004.01791v1 (accessed on 2 June 2020).
18. Pulido Rodríguez, C.; Villarejo Carballido, B.; Redondo-Sama, G.; Guo, M.; Ramis, M.; Flecha, R. False news around COVID-19 circulated less on Sina Weibo than on Twitter. How to overcome false information? *Int. Multidiscip. J. Soc. Sci.* **2020**, *1*. [CrossRef]
19. Barreneche, S.M. Somebody to blame: On the construction of the other in the context of the COVID-19 outbreak. *Soc. Regist.* **2020**, *4*, 19–32. [CrossRef]
20. Jarynowski, A.; Wójta-Kempa, M.; Płatek, D.; Czopek, K. Attempt to understand public-health relevant social dimensions of COVID-19 outbreak in Poland. *Soc. Regist.* **2020**, *4*, 7–44. [CrossRef]
21. Bikbov, B.; Bikbov, A. Communication on COVID-19 to community—Measures to prevent a second wave of epidemic. *OSF Preprints* **2020**. [CrossRef]
22. Kakodkar, P.; Kaka, N.; Baig, M. A comprehensive literature review on the clinical presentation, and management of the pandemic coronavirus disease 2019 (COVID-19). *Cureus* **2020**, *12*, e7560. [CrossRef]
23. Sá, M.J.; Serpa, S. The Global crisis brought about by SARS-CoV-2 and its impacts on education: An overview of the Portuguese Panorama. *Sci. Insights Educ. Front.* **2020**, *5*, 525–530. [CrossRef]
24. Douglas, M.; Katikireddi, V.; Taulbut, M.; McKee, M.; McCartney, G. Mitigating the wider health effects of covid-19 pandemic response. *BMJ* **2020**, *369*, m1557. [CrossRef] [PubMed]
25. Shuja, K.H.; Aqeel, M.; Jaffar, A.; Ahmed, A. COVID 19 Pandemic and impending global mental health implications. *Psychiatr. Danub.* **2020**, *32*, 1.
26. Gellert, G.A. Ethical imperatives critical to effective disease control in the coronavirus pandemic: Recognition of global health interdependence as a driver of health and social equity. *Online J. Health Ethics* **2020**, *16*. [CrossRef]
27. van Dorn, A.; Cooney, R.E.; Sabin, M.L. COVID-19 exacerbating inequalities in the US. *Lancet* **2020**, *395*, 1243–1244. [CrossRef]
28. Yancy, C.W. COVID-19 and African Americans. *JAMA* **2020**, *323*, 1891–1892. [CrossRef]
29. Costa, A.F. *Entrevista a António Firmino da Costa Sobre o Tema das Desigualdades Sociais. Um Olhar Sociológico Sobre a Crise Covid-19*; Associação Portuguesa de Sociologia: Lisbon, Portugal, 2020; Available online: youtu.be/XIoK7tsoZ7w (accessed on 25 June 2020).
30. Revet, S. *Covid-19: A Natural Disaster? Interview*; Center for International Studies: Paris, France, 2020.
31. Borraz, O. *Qu'Est-Ce Qu'Une Crise?* Centre de Sociologie des Organizations: Paris, France, 2020.
32. Dobry, M. *Sociologie des Crises Politiques: La Dynamique des Mobilisations Multisectorielles*; Presses de Sciences Po: Paris, France, 2009.
33. Older, M. Organizing after Disaster: The (re)Emergence of Organization within Government after Katrina (2005) and the Touhoku Tsunami (2011). Ph.D. Thesis, Institut d'études politiques de Paris, Paris, France, 2019.
34. Gilbert, C. *Power in Extreme Situations: Disasters and Politics*; L'Harmattan: Paris, France, 1992.
35. Lagadec, P. *States of Emergency: Technological Failures and Social Destabilization*; Éditions du Seuil: Paris, France, 1988.

36. Santos, B.S. *A Cruel Pedagogia do Vírus*; Coimbra [The cruel pedagogy of the virus]; Ediciones AKAL: Almedina, Portugal, 2020.
37. Zylberman, P. Health crises, political crises. *Trib. Santé* **2012**, *1*, 35–50. [CrossRef]
38. Ferreira, C.M.; Serpa, S. Special Issue "COVID-19 and Social Sciences". *Societies* **2020**. Available online: https://www.mdpi.com/journal/societies/special_issues/COVID-19_social_sciences (accessed on 2 June 2020).
39. Holmes, E.A.; O'Connor, R.C.; Perry, V.H.; Tracey, I.; Wessely, S.; Arseneault, L.; Bullmore, E. Multidisciplinary research priorities for the COVID-19 pandemic: A call for action for mental health science. *Lancet Psychiatr.* **2020**, *7*. [CrossRef]
40. Mogi, R.; Spijker, J. The influence of social and economic ties to the spread of COVID-19 in Europe. *SocArXiv Pap.* **2020**. [CrossRef]
41. van Bavel, J.J.; Baicker, K.; Boggio, P.S.; Capraro, V.; Cichocka, A.; Cikara, M.; Crockett, M.J.; Crum, A.J.; Douglas, K.M.; Druckman, J.N.; et al. Using social and behavioural science to support COVID-19 pandemic response. *Nat. Hum. Behav.* **2020**, *4*, 460–471. [CrossRef]
42. Costa, A.F. *Sociologia [Sociology]*; Difusão Cultural: Lisbon, Portugal, 1992.
43. Ferreira, C.M.; Serpa, S. Challenges in the Teaching of Sociology in Higher Education. Contributions to a Discussion. *Societies* **2017**, *7*, 30. [CrossRef]
44. Almeida, J.F. *Introdução à Sociologia*; Universidade Aberta: Lisboa, Portugal, 1994.
45. Elias, N. *The Society of Individuals*; Blackwell: Oxford, UK, 1991.
46. Flick, U. *Métodos Qualitativos na Investigação Científica [Qualitative Methods in Scientific Research]*; Monitor: Lisbon, Portugal, 2005.
47. Lalanda-Gonçalves, R. O documento nas ciências sociais: Construção e contextos sociais [The document in social sciences: Construction and social contexts]. In Proceedings of the 3rd International Colloquium of the MUSSI Network "The transformations of the document in the space-time of knowledge", São Salvador da Bahia, Brazil, 10–12 November 2014.
48. Carvalho, L.M. *Nós Através da Escrita: Revistas, Especialistas e Conhecimento Pedagógico (1920–1936)*; [Us through writing: Journals, specialists and pedagogical knowledge (1920–1936)]; Educa: Lisbon, Portugal, 2000.
49. Ramos, A.G. *The New Science of Organizations: The New Science of Organizations: A Reconceptualization of the Wealth of Nations*; University of Toronto Press: Toronto, ON, Canada, 1981.
50. Teixeira, R. Epidemic and security world. *Interface* **1998**, *2*, 77–96. [CrossRef]
51. Bourdelais, P. Contagions d'Hier et d'Aujourd'Hui [Contagions of yesterday and today]. *Soc. Sci. Health* **1989**, *7*, 7–20.
52. WHO. *Origin of SARS-CoV-2*; WHO: Geneva, Switzerland, 2020; Available online: https://www.who.int/publications/i/item/origin-of-sars-cov-2 (accessed on 26 June 2020).
53. Instituto de Higiene e Medicina Tropical. *Dossier: Origem e Dispersão Pandémica do Coronavírus SARS-CoV-2, Causador da COVID-19. IHMT*; Universidade Nova de Lisboa: Lisbon, Portugal, 2020.
54. Institut Pasteur. Covid-19 Disease (Novel Coronavirus). Available online: https://www.pasteur.fr/en/medical-center/disease-sheets/covid-19-disease-novel-coronavirus (accessed on 25 June 2020).
55. Rossman, J. Coronavirus: Can Herd Immunity Really Protect Us? Available online: https://www.weforum.org/agenda/2020/03/coronavirus-can-herd-immunity-really-protect-us (accessed on 30 June 2020).
56. Vilaça, O. Ce que les épidémies nous disent sur la mondialisation. Available online: http://cafe-geo.net/wp-content/uploads/epidemies-mondialisation.pdf (accessed on 25 June 2020).
57. Dollfus, O. *L'Espace Monde*; Economica: Paris, France, 1994.
58. Ludovic, J.; Bourdin, S.; Nadou, F.; Noiret, G. Economic globalization and the COVID-19 pandemic: Global spread and inequalities (Preprint). *Bull. World Health Organ.* **2020**. [CrossRef]
59. Correia, T. Entrevista a Tiago Correia Sobre o Tema da Saúde. Um Olhar Sociológico Sobre a Crise Covid-19. Associação Portuguesa de Sociologia. 2020. Available online: www.youtube.com/watch?v=SJPb_oW0kVg (accessed on 28 June 2020).
60. Beck, U. *World at Risk*; Polity Press: Cambridge, UK, 2008.
61. Mendes, J. Ulrich Beck: The immanence of social and the society of risk. *Soc. Anal.* **2015**, *214*, 2182–2999.
62. Poulain, M. *La Société Internationale Face au SARS: La Santé Publique a l'Épreuve de la Globalisation [International Society Facing SARS: Public Health Under the Globalisation Test]*; AFRi, Centre Thucydide—Analyse et Recherche en Relations Internationales, Université Paris II: Paris, France, 2005; Volume 6, pp. 905–921.

63. Bonnet, F.; Torny, D. The Very Thought of What Crisis Preparedness is Totally Absent. *Mediapart*. Available online: http://www.csi.mines-paristech.fr/en/featured-articles/the-very-thought-of-what-crisis-preparedness-is-is-totally-absent/ (accessed on 24 June 2020).
64. Lakoff, A. Preparing for the next emergency. *Public Cult.* **2006**, *19*, 247–271. [CrossRef]
65. Lakoff, A. The generic biothreat, or, how we became unprepared. *Cult. Anthropol.* **2008**, *23*, 399–428. [CrossRef]
66. Torny, D. From risk management to security production: The example of pandemic influenza preparedness. *Réseaux Découverte* **2012**, *30*, 45–66. [CrossRef]
67. Gilbert, C.; Raphaël, L. Towards political management of health crises? *Health Forums 3* **2011**, *2*, 55–60.
68. Leandro, M.; Leandro, A.; Nogueira, F. The trust in question. Skateboards of trust, the trust of the streaks in modern societies. *Soc. Rev. Fac. Let. Univ. Porto* **2011**, *11*, 215–232.
69. Pires, R. *Entrevista a Rui Pena Pires Sobre o Tema da Democracia e Desigualdades. Um Olhar Sociológico Sobre a Crise Covid-19*; Associação Portuguesa de Sociologia: Lisbon, Portugal, 2020.
70. Mateus, A. *The Economic Crisis of COVID-19: Facts and Perspectives, Challenges and Responses*; EY-Parthenon: London, UK, 2020.
71. IMF. *World Economic Outlook (April 2020)*; International Monetary Fond: Washington, DC, USA, 2020.
72. OECD. *Coronavirus: The World Economy at Risk*; OECD: Paris, France, 2020.
73. Minard, A.; Robert, A. Évolution microbienne et histoire humaine. Entretien avec Jared Diamond. *Tracés* **2011**, *21*, 217–234. [CrossRef]
74. Jared, D. *Inequality Among Societies. Essai sur l'Homme et l'Environnement dans l'Histoire [Inequality Among Societies. Essay on Man and the Environment in History]*; Gallimard: Paris, France, 2000.
75. Peretti-Watel, P.; Châteauneuf-Malclès, A. Sociology of risk and health crises: A perspective on the coronavirus pandemic. *Ressour. Sci. Econ. Soc.* **2020**.
76. Carapinheiro, G. Saúde e doença: Um programa crítico de sociologia da saúde [Health and disease: A Critical Program of Health Sociology]. *Sociol. OnLine* **2011**, *3*, 57–71.
77. Amekudzi-Kennedy, A.; Labi, S.; Woodall, B.; Marsden, G.; Grubert, E. Role of socially-equitable economic development in creating resilient and sustainable systems: COVID-19-related reflections. *Preprints* **2020**. [CrossRef]
78. Anderson, R.; Heesterbeek, H.; Klinkenberg, D.; Hollingsworth, T.D. How will country-based mitigation measures influence the course of the COVID-19 epidemic? *Lancet* **2020**, *395*, 931–934. [CrossRef]
79. Wilder-Smith, A.; Freedman, D.O. Isolation, quarantine, social distancing and community containment: Pivotal role for old-style public health measures in the novel coronavirus (2019-nCoV) outbreak. *J. Travel Med.* **2020**, *27*. [CrossRef] [PubMed]
80. Richardson, E.T. Pandemicity, COVID-19 and the limits of public health "science". *BMJ Glob. Health* **2020**, *5*, e002571. [CrossRef] [PubMed]
81. Schiariti, V. Les droits humains des enfants en situation de handicap en cas d'urgence sanitaire: Le défi du COVID-19 [The human rights of children with disabilities in health emergencies: The challenge of COVID-19]. *Dev. Med. Child. Neurol.* **2020**, *62*, E7. [CrossRef]
82. Lewnard, J.A.; Lo, N.C. Scientific and ethical basis for social-distancing interventions against COVID-19. *Lancet Infect. Dis.* **2020**, *20*. [CrossRef]
83. Rosenfeld, D.L.; Rothgerber, H.; Wilson, T. Politicizing the COVID-19 pandemic: Ideological differences in adherence to social distancing. *SocArXiv Pap.* **2020**. [CrossRef]
84. Silva, A.S.; Pinto, J.M. A global vision of social sciences. In *Metodologia das Ciências Sociais [Methodology of Social Sciences]*; Afrontamento: Porto, Portugal, 1986; pp. 9–27.
85. Fassin, D. *L'espace Politique de la Santé [The Political Space of Health]*; PUF: Paris, France, 1996.
86. Nunes, J.A. Saúde, direito à saúde e justiça sanitária [Health, right to health and health justice]. *Rev. Crítica Ciências Soc.* **2009**, *87*. [CrossRef]
87. Nentwich, M. Cyberscience: Modelling ICT-induced changes of the scholarly communication system. *Inf. Commun. Soc.* **2005**, *8*, 542–560. [CrossRef]
88. Cardoso, G.; Jacobetty, P.O. Que Significa Open Science? 2010. Available online: www.lini-research.org (accessed on 22 June 2020).

89. Borges, M.M. *Reflexos da Tecnologia Digital no Processo de Comunicação da Ciência. Una mirada a la Ciencia de la Información Desde los Nuevos Contextos Paradigmáticos de la Postmodernidad*; José Vicentini Jorente, M., Llanes Padrón, D., Eds.; Oficina Universitária; Cultura Acadêmica: São Paulo, Brazil, 2017; pp. 179–196.
90. Fassin, D. Santé globale, un nouveau concept? Quelques enseignements de l'épidémie à virus Ebola. *Méd. Sci.* **2015**, *31*, 5. [CrossRef] [PubMed]
91. Ferreira, C.M.; Serpa, S. Contagions and social sustainability: Domains, challenges and sanitary devices. *Preprints* **2020**. [CrossRef]
92. Santoro, P. Coronavirus: La sociedad frente al espejo [Coronavirus: Society in front of the mirror]. Available online: https://theconversation.com/coronavirus-la-sociedad-frente-al-espejo-133506 (accessed on 23 June 2020).
93. Portuguese Sociological Associati. *Um Olhar Sociológico Sobre a Crise Covid-19 [A Sociological Look at the Covid-19 Crisis A Sociological Look at the Covid-19 Crisis]*; APS: Lisbon, Portugal, 2020.
94. Holt-Lunstad, J.; Smith, T.; Baker, M.; Harris, T.; Stephenson, D. Loneliness and social isolation as risk factors for mortality: A meta-analytic review. *Perspect. Psychol. Sci.* **2015**, *10*, 227–237. [CrossRef]
95. Teuton, J. *Social Isolation and Loneliness in Scotland: A Review of Prevalence and Trends*; NHS Health Scotland: Glasgow, UK, 2018; Available online: http://www.healthscotland.scot/media/1712/social-isolation-and-loneliness-in-scotland-a-review-of-prevalence-and-trends.pdf (accessed on 4 June 2020).
96. Mardellat, V. Is It Immoral to Choose Which Lives to Save? 2020. Available online: http://cespra.ehess.fr/index.php (accessed on 6 June 2020).
97. Dubar, C. Le Pluralisme en Sociologie: Fondements, Limites, Enjeu [Pluralism in Sociology: Foundations, Limits, Stakes]. *Socio Logos* **2006**, *1*. [CrossRef]
98. Sethuraju, N.; Prew, P.; Abdi, A.; Pipkins, M. The consequences of teaching critical sociology on course evaluations. *SAGE Open* **2013**, *3*, 1–15. [CrossRef]
99. Morrison, A. A sociologist teaches history: Some epistemological and pedagogical reflections. *Educ. Stud.* **2017**, *53*, 233–246. [CrossRef]
100. Ferguson, S.J. The center does hold: The sociological literacy framework. *Teach. Sociol.* **2016**, *44*, 163–176. [CrossRef]
101. Elias, N. *Introdução à Sociologia [Introduction to Sociology]*; Edições 70: Lisboa, Portugal, 2008.
102. Vieira, M.M. Converter incrédulos: A sociologia na cidade das ciências duras [Converting skeptics: Sociology in the city of hard sciences]. In *Sociologia no Ensino Superior: Conteúdos, Práticas Pedagógicas e Investigação [Sociology in Higher Education: Contents, Pedagogical Practices and Research]. Proceedings of the Meeting Sociologia no Ensino Superior: Conteúdos, Práticas Pedagógicas e Investigação [Sociology in Higher Education: Contents, Pedagogical Practices and Research]*, Porto, Portugal, 6–7 December 2002; Gonçalves, C.M., Rodrigues, E., Azevedo, N., Eds.; School of Humanities of the University of Porto: Porto, Portugal, 2004.
103. Mills, C.W. *The Sociological Imagination*; Oxford University Press: Oxford, UK, 1959.
104. von Braun, J.; Zamagni, S.; Sorondo, M.S. The moment to see the poor. *Science* **2020**, *368*, 214. [CrossRef]
105. Lunn, P.D.; Belton, C.A.; Lavin, C.; McGowan, F.P.; Timmons, S.; Robertson, D.A. Using behavioral science to help fight the coronavirus. *J. Behav. Public Adm.* **2020**, *3*. [CrossRef]
106. Goffman, E. *Stigma. Notes on the Manipulation of Deteriorated Identity*, 1st ed.; Editora Guanabara: Rio de Janeiro, Brazil, 1963.
107. Gaudillière, J.-P.; Keck, F.; Rasmussen, A. *Viruses, Humans, Knowledge, Epidemics: The Social Construction of What?* L'École des Hautes Études en Sciences Sociales: Paris, France, 2020.
108. Aristovnik, A.; Ravšelj, D.; Umek, L. A Bibliometric Analysis of COVID-19 across Science and Social Science Research Landscape. *Preprints* **2020**. [CrossRef]

© 2020 by the authors. Licensee MDPI, Basel, Switzerland. This article is an open access article distributed under the terms and conditions of the Creative Commons Attribution (CC BY) license (http://creativecommons.org/licenses/by/4.0/).

Concept Paper

Responding to Social Distancing in Conducting Stakeholder Workshops in COVID-19 Era

Catherine Tobin *, Georgia Mavrommati and Juanita Urban-Rich

School for the Environment, University of Massachusetts Boston, Boston, MA 02125-3393, USA; Georgia.Mavrommati@umb.edu (G.M.); Juanita.Urban-Rich@umb.edu (J.U.-R.)
* Correspondence: Catherine.Tobin001@umb.edu; Tel.: +1-617-287-7440

Received: 27 August 2020; Accepted: 9 December 2020; Published: 13 December 2020

Abstract: In March 2020, COVID-19 disrupted global society. Impacts as a result of COVID-19 were seen in all industries, including higher education research, which was paused in order to accommodate newly imposed restrictions. Social science research, specifically stakeholder engagement research, was one area that was potentially impacted given its need for person-to-person interaction. Here, we describe how we successfully adjusted our stakeholder engagement methodology to accommodate for socially distant requirements. Initially, we planned to host in-person workshops to assess stakeholder perceptions of microplastics impacts on oysters in Boston Harbor and coastal Massachusetts using the deliberative multicriteria evaluation (DMCE) methodology. To transfer these workshops online, we used familiar, open-access platforms, Zoom and GoogleDrive, to enable dialogue among participants and evaluate preferences. While modifications to length (5 to 3 h) and order (participants were asked to watch expert videos before their participation date) of the workshop were necessary, most other elements of the methodology remained the same for the online format. The main element that was lacking was the in-person interactions. However, with video conferencing tools available, this element was not completely lost.

Keywords: deliberative valuation; informed decision making; public engagement; COVID-19; stakeholder engagement

---

1. Introduction—COVID-19 and Social Distancing

According to O'Steen and Perry [1], "a 'disaster' provides a form of societal shock which disrupts habitual, institutional patterns of behavior". Given the abruptness of the COVID-19 outbreak, Bonaccorsi et al. [2] suggests that COVID-19 could produce effects similar to those of a large-scale disaster. The abruptness of COVID-19 and the need to social distance and the move to remote working and learning is challenging teachers and researchers to find new or innovative methods to continue teaching and doing research, this especially true for social scientists that are frequently using and looking at social interactions.

While research is still being done on the impacts of COVID-19 on the education system, it is evident that COVID-19 altered the normal functioning of school systems at all levels [3]. Prevented from being able to meet for classes, meetings, etc., in person, faculty, staff and students found themselves in a new world order in which they had to adjust to adhere to new public health guidelines [4]. Higher education, specifically, faced many challenges; classes were disrupted, research was paused, and timelines altered. Wiggington et al. describes a world in which academic institutions, especially in regions where community transmission was severe (e.g., United States, Europe and China), had to halt all 'nonessential' on-site research activities (both in-lab and field-based research) abruptly [4]. Still, the greatest impact, most likely, will be seen at the early-career scientist level (including, masters and PhD students) [4]. Wiggington et al. purports that approximately 80% of on-site research was impacted at their affiliated institutions and that there will be long-term economic ramifications as a result [4]. In terms of the United States in 2018, higher education institutions accounted for approximately 13 percent of the money spent nationally on research and development, and approximately 50 percent of the money spent on basic research nationwide [4]. Altogether, COVID-19 emphasized the need for scenario planning and disaster preparedness in all sectors of society, including the research community, so that learning could still occur [4,5].

For social scientists, much of the research requires human interaction which is restricted or banned in the COVID-19 era. Engaging stakeholders into environmental decision making is essential for gaining better understanding of peoples' perceptions and designing solutions that can be implemented to solve the issue under consideration [6]. For example, evaluating social learning via stakeholder engagement workshops is being threatened during these times because effective engagement requires the interplay of participants through dialogue and transparency [7]. According to Reed et al., social learning is a process in which societal change for social-ecological systems occurs when people have the opportunity to exchange thoughts, ideas and values with their peers while also learning from others additional ways to perceive these systems [8]. The elicitation of shared values, specifically, is most effective via social interaction, open dialogue and social learning [9].

Future functioning of sustainable, long-term businesses requires the evaluation of social learning through the use of stakeholder engagement to understand and address the interests of employees, customers, suppliers and the community at large [7]. However, COVID-19 changes how these interactions can happen for the indefinite future. The University of North Carolina provided examples of engagement modes that could assist researchers working with stakeholders during these times [10]. Examples include the use of phone, email, snail mail, online file sharing, social media, and virtual meetings [10].

The impacts of COVID-19 can be seen in all sectors; however, the continuation of successful research and business practices is essential to moving the world forward. Given the unknowns, in terms of how long we will need to remain socially distant, it could take a significant amount of time before research institutions reach a new normal [4]. Additionally, given the duration of remote working and today's technology there is a question of how or if this will change back after COVID-19. Remote working and social distancing present challenges both in terms of data generation, career advancements, and global research overall, unless we learn to adapt [4]. While there are limitations to research based on the new work-from-home modality, Shelley-Egan acknowledges that this is also an opportunity for the research community to "imagine, and actively design, a future academic modus operandi characterized by a more sustainable, equitable and societally relevant research system" [11]. In other words, research (both in the sense of conducting it and participating in it) should not be constrained by geography [11].

This paper was created to showcase how social science research, specifically stakeholder engagement using the Deliberative Multicriteria Evaluation [DMCE] method, could still take place in a virtually distant society. To best of our knowledge the DMCE methodology has solely been employed during in-person workshops. Thus, this paper illustrates how the DMCE methodology can be adapted and move completely online when this is forced by external factors such as a pandemic. As Shelley and Egen emphasize, "the ontology of what it is to be a researcher and the temporalities, role identities, methodologies and epistemologies tied up with that will, most likely, adopt a different hue in—and beyond—this crisis" [11]. This paper is meant to highlight the resources that are available to do so seamlessly and encourage a new way of thinking when planning for stakeholder engagement.

2. Materials and Methods

2.1. Deliberative Valuation Process

In recent decades, there has been a shift towards using deliberative valuation in environmental policy analysis to improve public participation [12]. The deliberative valuation is a new valuation paradigm that calls participants to form values collectively and brings together stakeholders and scientists from various disciplines [13]. As discussed by Irvine et al., the generation of these values must happen through some form of community interaction; it cannot happen by researching individuals alone [9]. Thus, small groups of people representing the community at-large (termed 'mock citizen juries') are brought together to reach decisions on a given topic or policy based on mutual agreement [14]. In this sense, diverse views can be easily represented in the decision-making process and policy makers improve their understanding of peoples' perceptions. A broader understanding of environmental issues may increase the likelihood of conflict resolution in the early stages of environmental planning [15].

The basic idea behind this paradigm is that ecosystem services have the characteristics of common and public goods, and we need to assign values collectively. Essentially, deliberative valuation allows participants to exchange knowledge and evaluate public and common goods by developing mutual understanding through well-reasoned dialogue and deliberation [16]. In this sense, participants have 'equal bargaining power' [12]. Through the process of deliberation, participants are able to refine their personal beliefs and compromise value judgements with fellow participants [12]. The output of this process includes quantitative data in the form of willingness-to-pay assessments or collective ratings or rankings that are able to be compared across actions [17].

Scientists play an important role in this process by communicating to participants the science behind the environmental issue under consideration, and by providing their expertise whenever it is requested. Benefits associated with this process include (1) being able to integrate different sources of knowledge in decision making, (2) building social learning, and (3) eliciting social values vs. individual values [16]. A key component of the deliberative valuation method is participants collaborating in a group setting in the form of a 'citizens jury' to promote the common good which would benefit the community overall, not just individuals.

In order to implement this process, we used the deliberative multicriteria evaluation (DMCE) method. The DMCE method combines the benefits of multicriteria decision analysis with deliberation [13]. One of the advantages of the multicriteria decision analysis is the ability to compare actions that use different forms of measurements (e.g., number of permits vs. water column depth) or cannot be quantified in monetary terms [17]. In practice, attributes are identified and measured using easily understandable indicators by the general public and trade-off weights among these attributes are compared across actions [17]. Combining multicriteria decision analysis with deliberation results in stakeholders interacting to generate potential solutions for the issue under consideration and define areas of disagreement or agreement [6].

Figure 1 outlines the general steps taken to conduct research using the DMCE methodology. Here, we only present how COVID-19 affected the implementation of Step 4: Weighting of the indicator criteria. Steps 1–5, including results, will then be addressed and discussed in an additional, peer-reviewed paper.

Step 1: Stakeholder Recruitment

Step 2: Determination of overall objectives of the decision process

Step 3: Selection of the indicator criteria

Step 4: Weighting of the indicator criteria

Step 5: Evaluation of the indicator criteria

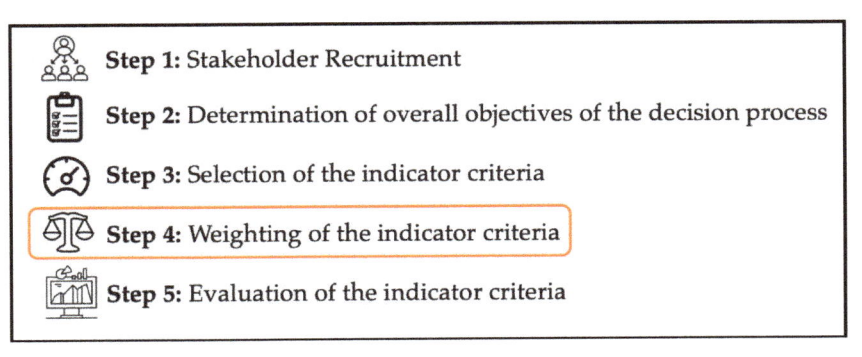

Figure 1. The steps of the deliberative process. This paper pertains to "Step 4: Weighting of the indicator criteria" which refers to conducting the stakeholder engagement workshops.

2.2. DMCE Methodology Example

Traditionally, when implementing the DCME methodology, stakeholders are asked to participate in an assessment task during an in-person workshop. These activities are dependent on the environmental issue being explored. For the case of this paper, we will provide a general example on how the DMCE methodology can be implemented using a car-buying analogy with the understanding that this typically happens through the convening of small groups. We will then discuss how this methodology was modified in order to be transitioned to a virtual format conducive to COVID-19 limitations.

When purchasing a new, used-car, there are some basic attributes that we could consider when shopping: mileage, miles-per gallon (MPG), and cost. If we put a 'value' of 'low' to 'high' on each, we are able to see the desired direction of each attribute (Table 1).

Table 1. Description of attributes.

Feature of Car	Measurement	Desired Direction	Scenario A (e.g., Worst Case Scenario)	Scenario B	Scenario C	Scenario D
Mileage	Low–High	Low	High	Low	High	High
Miles Per Gallon	Low–High	High	Low	Low	High	Low
Cost	Low–High	Low	High	High	High	Low

Since we do not want the worst-case scenario (indicated by 'Scenario A' in Table 1), we begin to populate the different scenario options in which one attribute is set at its 'best' value (cells highlighted in yellow in Table 1) while the others remain at their worst. These different scenarios are illustrated as cards (Figure 2) that participants are able to physically hold and compare visually to one another.

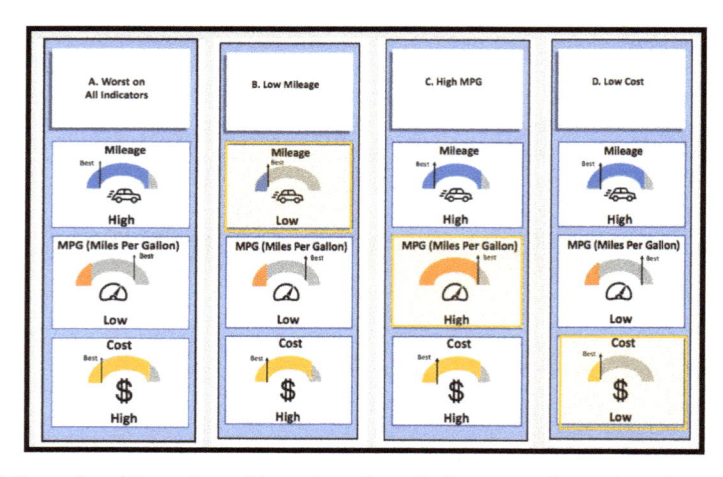

Figure 2. Examples of the cards used for each attribute. In these examples, each attribute is set to its worst level.

Participants are asked to prioritize the relative importance of these (4) attributes on a scale of 0–100 with 100 being the most preferred direction (Figure 3). Participants do these rankings as individuals at the beginning and end of the workshop and as a group throughout the workshop. Using the deliberative benefits of the DMCE methodology, the group is encouraged to discuss their value judgements, the reasoning behind their choices, compromise, and reach consensus within their stakeholder jury about the positioning and values of each scenario while collaborating as a group and relying on information from experts in the field that are provided to them during the workshop.

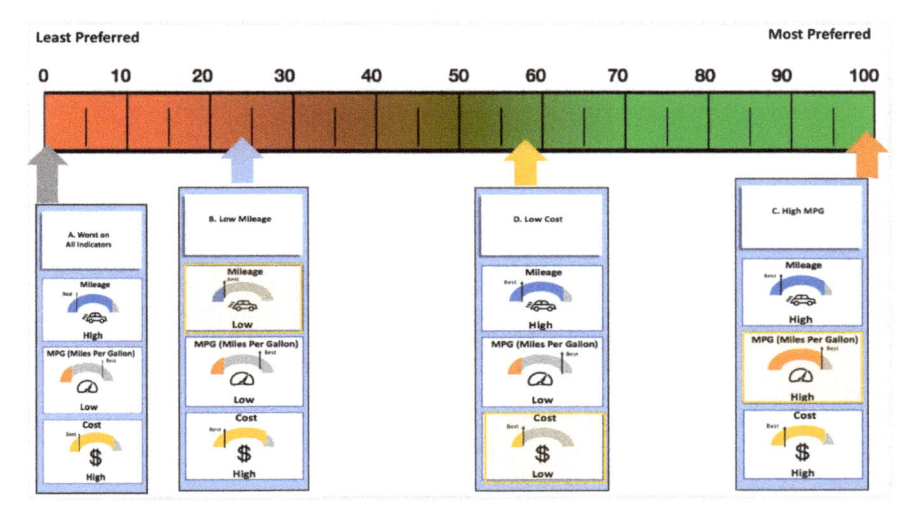

Figure 3. Example of the activity completed by participants. Scenarios (cards) are placed in the desired order on a scale of 0 (least preferred) to 100 (most preferred) based on individual and group preferences.

From the results of the choice tasks, trade-off weights for each attribute are calculated [16].

2.3. Initial Methodology Plans

The goal of this study was to assess stakeholder perception of microplastic pollution on oysters in Boston Harbor and coastal Massachusetts. Within this area there are different uses of oysters prevalent. Uses include oysters used as food; either grown in aquaculture, wild caught oysters or recreational fishing, and grown for environmental purposes (water clarity improvements, coastal erosion mitigation, etc.). This area is also a densely populated region in which microplastics are prevalent. With microplastic concentrations projected to increase in the next decade and an oyster's ability to ingest microplastics, Boston Harbor and coastal Massachusetts serve as relevant study areas to explore the interactions between these two actors.

Using the DMCE methodology, we initially planned to convene stakeholders in Massachusetts (including experts in the field related to oysters, shellfish, food security, marine debris, and microplastics) for a one-day, in-person workshop in April 2020. Stakeholders were contacted using the snowball sampling technique which consists of individuals from initial stakeholder categories identifying new stakeholders and contacts [18]. Stakeholders were from governmental agencies, academic institutions, non-governmental organizations, and industry. During the workshop, stakeholders were to be asked to assess the relative importance of socio-, economic, and environmental indicators, such as the attributes described for the car-buying analogy.

As of 20 March 2020, the University of Massachusetts Boston (the host institution) disbanded in-person classes, meetings and workshops due to COVID-19 related precautions. As such, the initial plans for implementing and executing this research needed to be addressed. Through a combination of timeline adjustments (switching from a full day workshop to a half day) and the use of familiar technology platforms (Zoom and GoogleDrive), we were able to successfully modify our methodology to accommodate stakeholder engagement during our new virtual reality.

3. Revisiting Our Methodology in a COVID-19 Era

To accommodate for the inability of large groups to gather, the implementation of this research had two options: (1) wait for a time in which large group gatherings were allowed or (2) modify the methodology and transition the workshops online. Given the uncertainty of the times, we opted to explore the second option.

In order to determine whether the second option was viable given the resources available, we decided to approach and address each of the following dimensions: technology, modifications (to methodology, agenda and IRB), and scheduling.

Technology: To acquire the appropriate data for this research, participants needed to be able to interact and 'move' the scenario cards based on their preferences. Additionally, we needed to be able to record the workshops. Thus, the ease of use, security and cost were of the utmost priorities when determining which platforms to use. Zoom, Microsoft Teams, GoogleDrive and Google Hangout, WebX, StarLeaf, GoToMeeting, iTricks and Digital Hives were platforms suggested by colleagues, with Zoom being the most common and inexpensive. In combination with Zoom, GoogleDrive would allow us to provide participants with individual and group links that they would then be able to manipulate in real time when necessary. A folder was created for each participant in GoogleDrive so that their data could be stored, and only participants and hosts had access to it.

Modifications to Methodology: Fortunately, through the use of Zoom and GoogleDrive, the methodology did not need to be modified drastically. We were still able to evaluate individual and group preferences using the scenario cards and group discussions were still able to be had. The tactile ('moving' printed out versions of the cards) and in-person components were impacted the most but corrected through the use of technology and a moderator.

Modifications to Agenda: Originally, the in-person workshop was scheduled for 5 h (plus travel time by participants). However, given other work obligations and childcare, participants expressed their concern about being able to participate for the entirety of the pre-planned agenda (even virtually).

While there are many benefits to the new work-from-home lifestyle that COVID-19 enforced (decreased commute time, flexible schedules, productivity gains, etc.) [19], we still had to be mindful of the negative aspects that truncated a participant's availability from all day to just a few hours in the wake of transitioning these workshops virtually. Research has shown that COVID-19 created additional challenges for parents of children, especially those children of a younger age who were unable to go to school or childcare and could not be unsupervised [20,21]. These challenges affected parents' ability to work-from-home undisturbed [19]. Many co-parents also had to make decisions about who could work when and who took care of the children throughout the day [20,21]. In an email correspondence with a stakeholder, she acknowledged that her participation was now limited to half-a-day because her and her husband were both working and sharing childcare responsibilities (C.Tobin. personal communication, 3 April 2020).

A 2018 study found that American adults logged approximately 11 h of screen time per day [22]. Sanudo et al. found smartphone use increased ~2 h per day during the lockdown [23]. Another study found that adults who spent more than six hours behind screens are more likely to suffer from moderate to severe depression [22]. Without the face-to-face meetings that oftentimes reinvigorate our attention spans, research has also shown the impacts of virtual meetings on people's level of tiredness [11]. Given that our workshops were at the beginning of the pandemic, participants may not have been as screen time- fatigued.

In addition to childcare challenges and increased screen-time usage, COVID-19 also promoted a more sedentary lifestyle, which can have detrimental effects on physical and mental health [19,23–25]. Recent research found that percentage of U.S. adults sitting more than 8 h per day increased 24% from pre- (16%) to during- (40%) confinement [23]. These factors (childcare, increased sedentary lifestyle, and increased screen time) could have discouraged stakeholders from wanting to participate in this type of virtual workshop for a full day. Aristovnik et al. acknowledges that the success of the new work-from-home environment cannot be dependent solely on ease of technology; other challenges can include lack of motivation and the need for improved self-discipline [26]. While the authors were specifically referencing student studying habits at home, those same challenges could affect non-students as well [26].

Shelley-Egan discusses how COVID-19 has disproved the need for academia (and other professions) to be physically in-person to conduct work [11]. Now, professionals have access to a plethora of online webinars and conferences that they otherwise may not have been able to attend in-person [11]. Thus, and in combination with the factors discussed above, the whole online world is competing against one another, vying for people's time. Since the data for this research was dependent on full participation, we had to adjust accordingly.

In order to truncate the agenda while not losing the necessary time for group deliberation, we decided to have experts pre-record their presentations that would provide context to what we would be discussing during the workshop. In doing so, we divided the workshop into three phases:

- Phase I occurred prior to the participation date. Participants were expected to watch the pre-recorded presentations from the experts;
- Phase II was the actual workshop on a given participation date; and
- Phase III occurred after the participation date. Participants were expected to complete a post-survey within one week.

Modifications to IRB: The Institutional Review Board (IRB) at UMB had previously reviewed and approved our research proposal in early winter 2020. At that time, the workshops were planned in-person (over 2 days) with 3 deliberative groups per day and did not have content pertaining to the use of Zoom and GoogleDrive. Thus, we needed to submit a modification to our research proposal that would allow us to conduct the research in this new manner. The modifications included changing the workshop to web-based using the online platforms of Zoom and GoogleDrive, increasing the number of workshops from 2 to 3 and decreasing the number of deliberative groups per day from 3 to 1. Originally, the average IRB review time during COVID-19 was 3 weeks, however, they were able to review and approve in one day. If IRB had not reviewed so quickly, we would have been tasked with moving the workshops further into the summer.

Scheduling of workshops: The in-person workshops were planned for the end of April 2020. On 20 March 2020, UMB suspended all in-person meetings, especially with large groups. While participants had already agreed to the in-person dates, we knew that it would take time to completely switch the workshops online. With that, we chose to push back the dates to the end of May 2020.

Scheduling for participants: Immediately upon learning about the impacts of COVID-19 to in-person stakeholder research, we contacted the participants and notified them of the postponement of the workshops with the understanding that we were exploring ways to host them virtually. Once we had identified new dates in late-May, we confirmed their willingness and ability to participate with the caveat that we were still modifying the agenda and determining which technology platforms to use. Details were shared with participants as they were approved (IRB) and confirmed (technology platforms). It was essential that we shared the technology platforms that were to be used ahead of time so we could identify participants who were unfamiliar with said platforms, unable to use them given their at-home workstations, or unable to use them given affiliation rules. We had one participant who was unable to participate given an agency-wide mandate on the use of Zoom.

4. Discussion

4.1. Opportunities in Virtual Stakeholder Engagement

Overall, hosting the stakeholder meetings virtually using Zoom and GoogleDrive was successful. After participating in the workshop, stakeholders had the opportunity to complete an evaluation form to let us know about their experience, responses included:

- "I appreciate how difficult this was over Zoom, how difficult surveys are in general and how difficult trying to get consensus on the topic of microplastic can be. I found the discussion interesting and informative for our own work communicating about this problem. I look forward to seeing the results!"

- "Getting to consensus with a group of wide-ranging stakeholder views is challenging. A well-executed virtual workshop. Thanks."
- "The facilitators and format was excellent, and the zoom and google docs worked really well. This is a hard kind of stakeholder prioritization because it is kind of theoretical or hypothetical and not so real life. But this is the kind of information you need to better understand stakeholder values."
- "Very impressed at the effort it took to organize this and provide such clear direction for us."
- "Good learning experience and networking opportunity."

Conducting in-person social science research can be costly in terms of funding and time. Expenses include room reservations (multiple if breakout groups are necessary), mileage reimbursement for participants, hotel reservations for experts, video-recording equipment, food, and supplies. Given the limited funding in social sciences, these factors can be a major deterrent from being able to conduct stakeholder workshops. Additionally, Shelley-Egen emphasizes that there is a suite of technological alternatives already in existence [11]. However, most of those alternatives were created in conjunction with face-to-face meetings [11]. Thus, a challenge for this research was to utilize platforms that would eliminate the need for in-person interactions.

We chose technology platforms that were not only familiar to participants but also at a low cost to us (GoogleDrive services were free and the upgraded Zoom account was already provided to professors from the University). There were no costs associated with on-boarding participants to these platforms since Zoom is free and GoogleDrive allows the host to send materials for review and manipulation by non-Gmail users. There were also no expenses associated with travel by hosts, experts, or participants. Oftentimes, the funding constrains the number of people who can participate given the associated costs per person. With this virtual workshop, we were able to invite as many stakeholders as appropriate. Additionally, over the three days, only 10 percent of participants recruited did not show up, whereas 30 percent of participants recruited for previous work did not show up [16]. Altogether, using virtual platforms to conduct social science research removes the constraints detailed above and may provide an opportunity for the direction of future social science research.

We acknowledge that the success of this methodology is constrained by internet connectivity and quality and users comfort with technology and remote meetings. Belzunegui-Eraso and Erro-Garces emphasize the important role technology plays in the successful development of telework [27]. According to Belzunegui-Eraso and Erro-Garces, the ideal teleworker is a "Millenial woman holding a higher education degree, with 4–10 years of professional experience, and working from home two days a week in the management and administration field" whereas the least ideal person is a "man of the baby boomers' generation, holding a university degree, with 20 years or more of professional experience, and who started working remotely only during quarantine" [27]. However, conducting human-centric research does not always allow for the ideal teleworker. Thus, this [meaning, technology literacy] is an important indicator of successful research during these COVID-19 times; future work should keep in mind socioeconomic status and geographic location of participants to assure that data is irrespective of technological challenges. For our research, we assumed that the participants were technologically literate, given their professions and backgrounds, but we also provided an opportunity prior to the workshop to confirm and adjust based on their technological literacy as needed.

Participants were required to view presentations ahead of their participation date. The presentations were from experts in the field of microplastics and oysters as well as one from the host explaining the general agenda of the workshop as well as the indicators that were to be evaluated by participants day-of. Typically, these presentations would have been done at the beginning of the workshops; however, in the initial discussions with participants about changing the format of the workshops from in-person to online, participants expressed that their time availability had become truncated, in essence due to other work obligations and childcare. Taking that into account, we shortened the time of the workshop from 5 h to 3 h and moved certain agenda items (e.g., expert presentations) prior to the workshop to be sure that participants had enough time for group deliberation. As a result of this, we may have encouraged more folks to participate given that they did not have to travel to Boston in order to participate (as was the plan during initial outreach). As of March 2020, we had 20 stakeholders agree to participate in the in-person workshop (30% from the Academic sector, 30% from Government, 25% from NGO, and 15% from Industry). At the completion of the virtual workshops, 30 stakeholders had participated (21% from the Academic sector, 41% from government, 24% from NGO, and 14% from Industry). This increase in stakeholders could be attributed to increased outreach and the virtual nature of the workshops. However, given childcare constraints, as discussed in the previous section, some may have still been discouraged to participate.

The change from in-person to virtual was communicated via email through the description of the three participation phases:

- Phase I took place before the participation date. Participants were asked to watch four presentations (totaling, approximately 40 min). One provided context to the workshop activities and the other three were from three experts in the field of shellfish and microplastics. The content in these presentations provided the necessary background information for the workshop.
- Phase II took place on the respective participation date. Participants needed the (2) weblinks for Google Slides and Zoom.
- Phase III took place 1 week after the workshop. Participants were asked to complete an activity similar to what they did day-of (no longer than 10–15 min). Participants completed this survey in their individual GoogleSlides link.

The expert presentations were uploaded to a shared GoogleFolder and shared with participants 5 days prior to their participation date. One participant said "I liked participating, it was very interesting and the format was fine, nice job! I am happy to have been part of this. I liked being able to listen to the information ahead of time and to hear a range of comments from such a group of stakeholders … " When asked if participants were able to successfully view the presentations, 83 percent of participants indicated 'yes,' 10 percent indicated 'partially' and 7 percent indicated 'no' (Figure 4A). All participants found the instructional material provided was helpful either in totality (67 percent) or partially (33 percent) (Figure 4B). Additionally, participants were exposed to the key points of each experts' pre-workshop presentation at the start of Phase II; participants were offered the opportunity to ask questions and/or clarify points made by the experts at this time. For future work, one participant suggested that the videos be placed on a secure Youtube Channel rather than GoogleDrive. Furthermore, click rates could be monitored to assure that the videos were actually viewed ahead of the workshops and an additional question could be added to the post-workshop evaluation form asking participants if they viewed the material undistracted or while doing other work.

When these workshops are video-recorded in-person, it is evident to participants that a camera is recording their conversations and movements. This may become a hindrance to 'camera shy' individuals. With this new virtual format, participants were read a consent form pertaining to the use of video-recording at the start of the workshop. The camera was not as apparent throughout the virtual workshop (as it would have been in-person) which may have allowed those individuals to participate more comfortably.

We also decided to keep the daily groupings smaller in number compared to doing it in-person. Acknowledging that free-flowing conversation is not as easy in a virtual workshop, we wanted to make sure that all voices could be heard. Aided by a facilitator, participants were called on to help move the conversation forward when needed. One participant mentioned that they "appreciated the moderator calling on people to speak up. Often one person dominates the conversations, especially during remote meetings … Getting more people to speak can only improve the discussion … ". Even with the virtual format, however, participants were generally satisfied with the outcome of the group deliberation (Figure 4C).

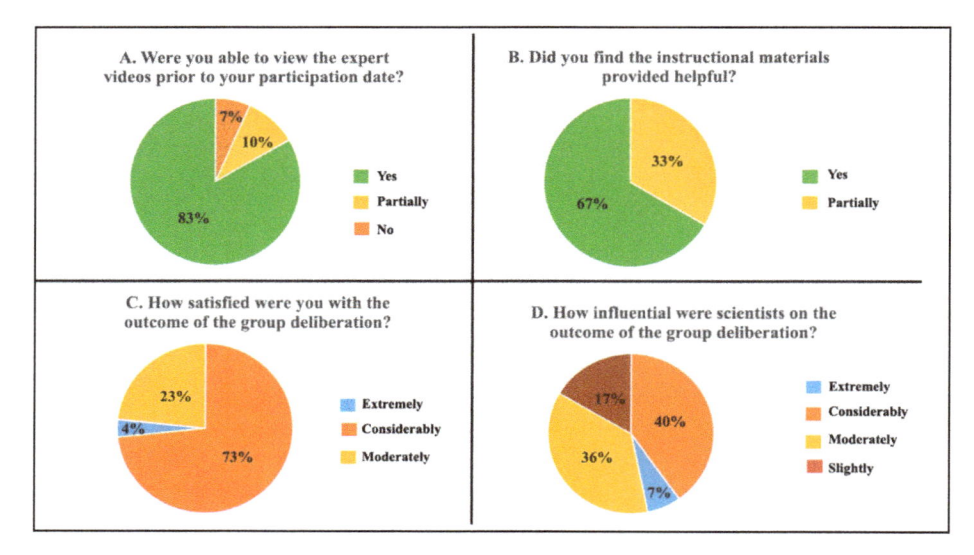

Figure 4. Responses from the evaluation form that participants were asked to complete post-workshop.

4.2. Challenges in Virtual Stakeholder Engagement

At the crux of stakeholder engagement, and specifically the DMCE methodology, is the ability to interact with fellow participants. This interaction is important for both the researchers and the participants. For the researchers, the qualitative data (e.g., the discussions between participants) are just as, if not more important, than the quantitative data (e.g., the ranking values given to the scenarios). We wanted to understand the reasoning behind choices (why did participants preferences change during the group deliberation when compared to pre-deliberation preferences?). For the participants, being able to learn from other participants during the group deliberation is essential to the learning process that takes place while implementing the DMCE methodology. While using Zoom was a decent substitute for in-person relations (one participant said "We are all becoming more comfortable and competent with Zoom. It will be a most valuable tool in the future."), not all participants were able to use Zoom as a result of work-related mandates (one participant, who had initially agreed to participate, had to cancel because of the use of Zoom). Since participant and data security is integral to working with human subjects, using platforms that provide that is essential. Based on this experience, we can identify how we need to improve technological tools that we used in this process.

Lacking these in-person relations was one of the major challenges of this research. Previous work done using this methodology emphasized the need for the in-person relations [16]. Typically, participants would have been provided with a printout of the scenario cards in addition to the scale bar so that individuals could move the cards based on their preferences in real-time and others could evaluate and suggest changes, thus prompting conversation among participants. In hopes of providing some of that tactile learning, we provided images of the cards on a GoogleSlide (that all participants had access to) so that they could move the cards in real-time. What we found, however, is that most participants did not utilize that function during the group deliberation section of the workshop and thus we required a graduate student to help move the cards based on the discussion.

One of the unique aspects of this methodology is being able to evaluate participant preferences before, during, and after knowledge consumption in order to see if preferences change or stagnate with the gaining of more knowledge. Typically, all of the information would have been provided day-of so that we were able to record initial preferences in real-time, rather than allow the participants to contemplate the information for days prior. However, as mentioned above, we had to adapt to current times by shortening the day-of workshop and providing materials ahead of time. Given that materials were provided 5 days in advance, participants technically had the opportunity to reflect on the content, specifically the indicators, prior to completing the initial, individual survey day-of. For future work, participants could be asked to complete the initial, individual survey when they received the materials in advance.

Providing opportunities for participants to gain knowledge throughout the workshop is integral to successfully implementing this methodology. Knowledge building can happen during structured expert presentations, conversations with other participants, as well as candid conversations with experts throughout the day. For the latter, participants are told that experts are available throughout the workshop to clarify any points/questions. While experts were present during each virtual workshop for its entirety, participants oftentimes forgot they were there since they (the experts) are not meant to interject unless called upon, which may have led to the discrepancies in how influential the scientists ('experts') were on the outcome of the group deliberation (Figure 4D).

5. Conclusions

COVID-19 has challenged society's social abilities. With society's new constraints, academic research had to adapt in order to make progress. Social science research, specifically stakeholder engagement, was one such area that met many limitations during the pandemic. The deliberative multicriteria evaluation process builds upon a methodology in which human interaction is essential to the outcomes of the process. Here, we discussed how we were able to modify the DMCE methodology so that the experiment could still take place during a COVID-19 era with no in-person interaction allowed. To do so, we built upon the principles of the DMCE methodology and transferred it online using resources that were publicly available, free, and already used by the participants. While modifications to length (5 to 3 h) and order (participants were asked to watch expert videos before their participation date) of the workshop were necessary, most other elements of the methodology remained the same for the online format. The main element that was lacking was the in-person interactions. However, with video conferencing tools available, this element was not completely lost. In agreement with the findings of Liu et al., our research highlighted the use and easy deployment of education technology to expand traditional social science research and engage with a more diverse (in terms of relationship to oysters and microplastics in Massachusetts) audience [28].

Society, including the research community, may interact very differently in a post-COVID-19 world. This paper highlights how stakeholder engagement can still occur in this new virtual reality. Future research can explore how to improve and encourage the discourse between participants using video conferencing tools (e.g., through the use of breakout rooms, better facilitation techniques). While keeping costs low was an aim of this research, exploring other platforms that are not free may strengthen the outcomes of the workshops as those platforms may have better tools to engage with and between participants.

Author Contributions: Conception and design: C.T., G.M., J.U.-R. acquisition of data: C.T., G.M., J.U.-R. and interpretation of data: C.T., G.M. Drafted and/or revised the article: C.T., G.M., J.U.-R. Approved the submitted version for publication: C.T., G.M., J.U.-R. All authors have read and agreed to the published version of the manuscript.

Funding: This work is supported the National Science Foundation IGERT: Coasts and Communities—Natural and Human Systems in Urbanizing Environments Program NSF DGE 1249946.

Acknowledgments: The stakeholder workshops were approved by the University of Boston IRB # 2020013. We thank Juanita Urban-Rich, Evan J. Ward, and Abigail Barrows for serving as experts, Jennifer Bender for serving as the facilitator, and all of the stakeholders who participated.

Conflicts of Interest: The authors declare no conflict of interest.

References

1. O'Steen, B.; Perry, L. Service-Learning as a Responsive and Engaging Curriculum: A Higher Education Institution's Response to Natural Disaster. *Curric. Matters* **2012**, *8*, 171. [CrossRef]
2. Bonaccorsi, G.; Pierri, F.; Cinelli, M.; Flori, A.; Galeazzi, A.; Porcelli, F.; Schmidt, A.L.; Valensise, C.M.; Scala, A.; Quattrociocchi, W.; et al. Economic and social consequences of human mobility restrictions under COVID-19. *Proc. Natl. Acad. Sci. USA* **2020**, *117*, 15530–15535. [CrossRef] [PubMed]
3. Toquero, C.M. Challenges and Opportunities for Higher Education amid the COVID-19 Pandemic: The Philippine Context. *Pedagog. Res.* **2020**, *5*, em0063. [CrossRef]
4. Wigginton, N.S.; Cunningham, R.M.; Katz, R.H.; Lidstrom, M.E.; Moler, K.A.; Wirtz, D.; Zuber, M.T. Moving academic research forward during COVID-19. *Science* **2020**, *368*, 1190–1192. [CrossRef]
5. Yan, W. Early-career scientists at critical career junctures brace for impact of COVID-19. *Science* **2020**. [CrossRef]
6. Gregory, R.; Failing, L.; Harstone, M.; Long, G.; McDaniels, T.; Ohlson, D. *Structured Decision Making: A Practical Guide to Environmental Management Choices*; John Wiley & Sons: Hoboken, NJ, USA, 2012.
7. Michael, M. The importance of stakeholder engagement during the coronavirus pandemic. 2020. Available online: https://www.openaccessgovernment.org/stakeholder-engagement-during-coronavirus/85915/ (accessed on 2 November 2020).
8. Reed, M.S.; Evely, A.C.; Cundill, G.; Fazey, I.; Glass, J.; Laing, A.; Newig, J.; Parrish, B.; Prell, C.; Raymond, C.; et al. What is social learning? *Ecol. Society* **2010**, *15*. [CrossRef]
9. Irvine, K.N.; O'Brien, L.; Ravenscroft, N.; Cooper, N.; Everard, M.; Fazey, I.; Reed, M.S.; Kenter, J.O. Ecosystem services and the idea of shared values. *Ecosyst. Serv.* **2016**, *21*, 184–193. [CrossRef]
10. Community, T. *Engaged Research Partnerships during COVID-19: Staying True to Principles of Community and Stakeholder Engagement during the COVID-19 Pandemic*. pp. 1–9. Available online: www.ARCCresources.net (accessed on 5 November 2020).
11. Shelley-Egan, C. Testing the obligations of presence in academia in the COVID-19 era. *Sustainability* **2020**, *12*, 6350. [CrossRef]
12. Howarth, R.B.; Wilson, M.A. A theoretical approach to deliberative valuation: Aggregation by mutual consent. *Land Econ.* **2006**, *82*, 1–16. [CrossRef]
13. Raymond, C.M.; Kenter, J.O.; Plieninger, T.; Turner, N.J.; Alexander, K.A. Comparing instrumental and deliberative paradigms underpinning the assessment of social values for cultural ecosystem services. *Ecol. Econ.* **2014**, *107*, 145–156. [CrossRef]
14. Proctor, W.; Drechsler, M. Deliberative multicriteria evaluation. *Environ. Plan. C* **2006**, *24*, 169. [CrossRef]
15. Mavrommati, G.; Rogers, S.; Howarth, R.B.; Borsuk, M.E. Representing Future Generations in the Deliberative Valuation of Ecosystem Services. *Elementa* **2020**, *8*, 22. [CrossRef]
16. Mavrommati, G.; Borsuk, M.E.; Howarth, R.B. A novel deliberative multicriteria evaluation approach to ecosystem service valuation. *Ecol. Soc.* **2017**, *22*. [CrossRef]
17. Borsuk, M.E.; Mavrommati, G.; Samal, N.R.; Zuidema, S.; Wollheim, W.; Rogers, S.H.; Thorn, A.M.; Lutz, D.; Mineau, M.; Grimm, C.; et al. Deliberative multiattribute valuation of ecosystem services across a range of regional land-use, socioeconomic, and climate scenarios for the upper merrimack river watershed, New Hampshire, USA. *Ecol. Soc.* **2019**, *24*. [CrossRef]
18. Naderifar, M.; Goli, H.; Ghaljaie, F. Strides in development of medical education. *Strides Dev. Med. Educ.* **2017**, *14*, 1–6. [CrossRef]
19. Kaushik, M.; Guleria, N. The Impact of Pandemic COVID -19 in Workplace. *Eur. J. Bus. Manag.* **2020**, 8–18. [CrossRef]
20. Qian, Y.; Fuller, S. COVID-19 and the gender employment gap among parents of young Children. *Can. Public Policy* **2020**, *46*, S89–S101. [CrossRef]
21. Salin, M.; Kaittila, A.; Hakovirta, M.; Anttila, M. Family coping strategies during finland's COVID-19 lockdown. *Sustainability* **2020**, *12*, 9133. [CrossRef]
22. Akulwar-Tajane, I.; Parmar, K.K.; Naik, P.H.; Shah, A.V. Rethinking Screen Time during COVID-19: Impact on Psychological Well-Being in Physiotherapy Students. *Int. J. Clin. Exp. Med. Res.* **2020**, *4*, 201–216. [CrossRef]
23. Sañudo, B.; Fennell, C.; Sánchez-Oliver, A.J. Objectively-assessed physical activity, sedentary behavior, smartphone use, and sleep patterns preand during-COVID-19 quarantine in young adults from Spain. *Sustainability* **2020**, *12*, 5890. [CrossRef]

24. Colley, R.C.; Bushnik, T.; Langlois, K. Exercise and screen time during the COVID-19 pandemic. *Health Rep.* **2020**, *31*, 1–11. [CrossRef]
25. Sultana, A.; Tasnim, S.; Bhattacharya, S.; Hossain, M. *Digital Screen Time during COVID-19 Pandemic: A Public Health Concern*. 2020, pp. 1–14. Available online: https://osf.io/preprints/socarxiv/e8sg7/ (accessed on 5 November 2020).
26. Aristovnik, A.; Keržič, D.; Ravšelj, D.; Tomaževič, N.; Umek, L. Impacts of the COVID-19 pandemic on life of higher education students: A global perspective. *Sustainability* **2020**, *12*, 8438. [CrossRef]
27. Belzunegui-Eraso, A.; Erro-Garcés, A. Teleworking in the context of the Covid-19 crisis. *Sustainability* **2020**, *12*, 3662. [CrossRef]
28. Liu, Y.; Zhang, Y.; Qiao, W.; Zhou, L.; Coates, H. Ensuring the sustainability of university learning: Case study of a leading Chinese University. *Sustainability* **2020**, *12*, 6929. [CrossRef]

Publisher's Note: MDPI stays neutral with regard to jurisdictional claims in published maps and institutional affiliations.

© 2020 by the authors. Licensee MDPI, Basel, Switzerland. This article is an open access article distributed under the terms and conditions of the Creative Commons Attribution (CC BY) license (http://creativecommons.org/licenses/by/4.0/).

Article

Altered Self-Observations, Unclear Risk Perceptions and Changes in Relational Everyday Life: A Qualitative Study of Psychosocial Life with Diabetes during the COVID-19 Lockdown

Dan Grabowski *, Julie Meldgaard and Morten Hulvej Rod

Steno Diabetes Center Copenhagen, Niels Steensens Vej 6, 2820 Gentofte, Denmark; julie.meldgaard.petersen@regionh.dk (J.M.); morten.hulvej.rod@regionh.dk (M.H.R.)
* Correspondence: dan.grabowski@regionh.dk

Received: 18 July 2020; Accepted: 30 August 2020; Published: 1 September 2020

Abstract: When the Danish society went into COVID-19 lockdown, it dramatically changed the conditions for living with a chronic disease like diabetes. The present article highlights the psychosocial effects of this change. The dataset consists of 20 semi-structured online interviews with people with diabetes. The data were analyzed using radical hermeneutics and interpreted using Luhmann's operative constructivist systems theory. The analysis produced three main themes: (1) people with diabetes experience altered self-observations–mainly due to society labelling them as vulnerable, (2) people with diabetes have unclear risk perceptions due to lack of concrete knowledge about the association between COVID-19 and diabetes, and (3) changes in conditions for maintaining and creating meaningful relations have a significant impact on everyday life with diabetes. These findings have important implications for risk communication. People respond in a multitude of ways to communications issued by health authorities and with close relations, and their meaning-making is shaped by, and shapes, their self-observations, risk perceptions and relational environments. This calls for more targeted communication strategies as well as increased use of peer support; the goal being to help people create meaning in their own environments.

Keywords: COVID-19; diabetes; psychosocial effects; self-observations; risk perceptions; social relations; systems theory; qualitative research

---

1. Introduction

The spread of the 2019 novel coronavirus was declared a global pandemic by the World Health Organization on 11 March 2020 [1]. The virus most often only causes mild symptoms similar to those of a typical influenza, but in some people, it can cause severe respiratory infections and multi-organ failure [2]. These severe symptoms are more likely to occur in specific high-risk groups. The designation of such high-risk groups has evolved during the course of the pandemic and has differed across countries. On 11 March, the Danish Health Authority declared people with poorly regulated diabetes (all types) as a group at high risk for becoming seriously ill if infected with the virus [3].

Studies investigating the prevalence of diabetes (all types) in COVID-19 patients have found it to be around 10% [4,5], but it remains unclear whether people with diabetes are more susceptible to COVID-19 [6,7]. However, numerous studies have shown an increased risk of severe COVID-19 and mortality in individuals with diabetes [4,6–12], as well as higher rates of hospitalization due to COVID-19 [13,14]. Thus, individuals with diabetes have been identified as having worse COVID-19 outcomes, particularly those with poor glycemic control [15,16].

This information on the association between diabetes and COVID-19 was not, however, available during the early stages of the pandemic due to the novelty of the virus [17]. As a result of this uncertainty and lack of evidence-based knowledge about the virus, the information and messages from health authorities, and therefore also from the media, have been mixed and unclear, with different predictions and inconsistent risk communication [18]. This also applies to Denmark, where information from health authorities has varied over the course of the pandemic: Initially, only individuals with poorly regulated diabetes were said to be at risk of severe COVID-19, but later all people with diabetes, regardless of glycemic control and diabetes type, were categorized as a risk group [3].

Just as in many other countries, the Danish government enforced a lockdown to avoid critical spreading of the 2019 novel coronavirus [19]. On 12 March, Denmark closed all schools and childcare institutions, sent home public sector employees with non-critical roles and banned gatherings of more than 10 people.

A pandemic-induced lockdown can be expected to reveal social, psychological and underlying philosophical issues that will have a lasting impact on societies and individuals [20]. To study how a complex societal development such as the lockdown affects the self-observations and perceptions of communicated risk among people living with a chronic disease, we need to apply complexity-oriented social theory that enables us to look closely at the interrelatedness of the individual and the social. In the present study, we look through the lens of Niklas Luhmann's operative constructivist systems theory [21–23]. The objective of the present paper is to study how COVID-19 lockdown affects psychosocial life with diabetes. By applying significant concepts from Luhmann's systems theory, and furthermore by analyzing using radical hermeneutics [24], we examine how people with diabetes observe themselves in relation to other people, various social systems in their environment as well as the lockdown as an important societal phenomenon. Applying a qualitative approach with a distinct focus on individuals' thoughts, perceptions, and views enable us to understand how people with diabetes create meaning out of the complex multitude of communications concerning the consequences of the 2019 novel coronavirus.

2. Materials and Methods

2.1. Empirical Data

The dataset consists of 20 semi-structured individual interviews with people with diabetes (see Table 1 for characteristics). Due to the lockdown and the need for social distancing, the interviews were conducted and recorded online using a secure communication platform with the video function disabled. We decided to disable the video function because especially the older interviewees might be fairly unfamiliar with video conferences and could find this to be intimidating.

Table 1. Study population characteristics.

Participants	(n = 20)
Mean age in years, [range]	51.7 [20–75]
Gender, female	10
Diabetes type	
Type 1	8
Type 2	11
LADA *	1
Mean diabetes duration in years [range]	14.9 [0–48]

* Relatively rare form of diabetes.

The relatively simple interview guide consisted of four main themes: (1) diabetes management (in general and during lockdown): with questions focusing on the day-to-day issues regarding medication, exercise and diet, (2) quality of life (in general and during lockdown): with questions focusing on perceptions of general well-being, (3) perception of being at risk (in general and during lockdown): with questions focusing on health and risk communication and the specific relation between diabetes and COVID-19, and (4) relations and support (in general and during lockdown): with questions focusing on interactions with friends, family members, colleagues and healthcare professionals.

The participants were recruited through the user panels at Steno Diabetes Center Copenhagen and the Danish Diabetes Association. The user panels consist of people with diabetes who have volunteered to share information about their lives with diabetes. The panels include people from all parts of Denmark with type 1 diabetes (T1D) and type 2 diabetes (T2D, LADA, gestational diabetes and other rarer types of diabetes. Thus, the panels represent people with diabetes who are treated in different primary and/or secondary care settings across Denmark. The panel members ($n = 2430$) received the invitation to participate in an interview via an e-mail newsletter sent out to all panel members. The interested panel members then contacted the research team by e-mail or telephone.

We decided to be open to recruit participants with T1Dand T2D so that we could study potentially different reactions to the lockdown. During one interview we discovered that the interviewee actually had LADA, which is a relatively rare form of diabetes that has similarities with both T1D and T2D [25]. We decided to keep this interview in the dataset even though it did not contribute any insights into any specificities regarding living with LADA during lockdown. It did, however, contribute to the overall findings on living with diabetes during lockdown. Thus, in the present paper, we focus on persons with T1D and T2D

After each interview, the researchers discussed and systematically wrote down all the overarching themes that were touched upon in the particular interview. After 15 interviews, the research group used this overview to discuss the level of saturation and decided to conduct five additional interviews, because the interviews, especially with people with T2D, lacked information power. The recruitment procedure was not changed for the next five interviews that included three people with T2D and two people with T1D. When a total of 20 interviews had been conducted, we concluded that no new overarching new themes had emerged in the two latest interviews, meaning that the level of saturation and ensuing information power in the data were acceptable [26]. No changes were made in the interview guide at any point in the data collection period. All interviews were conducted over the course of two weeks and the average length was 40 min. None of the researchers had any prior relationship with any of the interviewees.

After collecting the full dataset, the interviews were transcribed verbatim, iteratively analyzed in Danish and then categorized using radical hermeneutics, which are a set of guidelines for content analysis that, as a combination of hermeneutics and constructivism, manages to simultaneously be empirically founded and theoretically complex [24]. Radical hermeneutics focuses on keeping a perpetual balance between theory, method and data by acknowledging how all of these elements influence each other in an interconnected process. The use of radical hermeneutics also entails constant alternation between analyzing and interpreting, which means it is necessary to present interpretive aspects while presenting the results.

Radical hermeneutics is a validated methodology consisting of three steps of data analysis. The first step involves reading the data with a view to observing specifically selected differences in them. This observation in itself constitutes an interpretation rather than a description, and its task is to reduce the complexity of the data. Elements within the scope of the differences selected by the interpreter are extracted from the data. The second step involves making these elements the subject of interpretation as an observation of the differences employed. The third step involves interpreting the sum of these differences [24].

In the present analysis, this approach meant that the analytical process was developed in several steps as we delved deeper into the data. The first step focused on extracting issues of direct relevance to diabetes and the COVID-19 lockdown from the empirical data. The second step involved analyzing and interpreting the extracted data using the theoretical background described below, and this step revealed the three main themes presented in the results section. The final step is then a separate interpretation of the data within each category–materializing as the findings presented for each of the three themes. Working with radical hermeneutics includes the awareness that the role of the researcher is that of an observer. This underlines how any research depends on an awareness of how all the observations were made. This is exactly what radical hermeneutics and the use of guiding differences provide—a way to continually employ a high level of reflexivity in all aspects of the study.

The study was approved by the Danish Data Protection Agency (P-2020-271) and carried out in accordance with the Declaration of Helsinki. According to Danish legislation, interview studies require no approval from an ethics committee. All participants gave their informed consent based on thorough explanations of the purpose of the study. All participants were explained that they had the right to not answer any question and that they could stop the interview at any time if they were not comfortable with the situation. Furthermore, they were given details of whom to contact for answers to questions about the research and the research subject's rights.

2.2. Theory

In the present study, we adopt the theoretical position that every person observing his or her own unique environment is observing something slightly different from what everyone else is observing. This is important to bear in mind in relation to COVID-19 risk communication, as every person will be observing this in a slightly different way, and the messages will therefore mean something different in the context of different environments [23].

Luhmann's systems theory is, in essence, a grand theory about communication and the individual person's observation and meaning-making of this communication. The theory introduces a distinction between psychological and social systems, which is an important aspect of our present approach to studying life with diabetes during lockdown. What Luhmann called psychological systems, in fact, constitute the consciousness of persons, individuals or subjects, and for Luhmann, social systems can be either interactions, organizations or societies [23].

The consciousness of psychological systems and the communication of social systems do not have direct access to each other. They basically speak different languages, and this means that the psychological systems cannot communicate, but are instead restricted to observing the communicative social systems. The systems are closed and autopoietic, which means that each system continually self-interprets and self-reflects on the basis of its own contents. This inability to fully understand the other kind of system is Luhmann's way of addressing the classical sociological issue of the co-evolution of the social and the individual [22].

The notion that the systems are closed means that the consciousness that continually recreates the way we (as psychological systems) interpret and understand ourselves, in the context of our own unique environment, always comes from within the system itself. Consciousness never comes from the outside, because we never acquire consciousness from other psychological or social systems. We are therefore closed in terms of this operational element. It is this self-reference that enables systems to continually distinguish themselves from their environment. The concept of self-reference is therefore essential in understanding how each system is able to maintain its own horizon of meaning despite the complexity of the environment [21].

The psychological system continues to produce the consciousness needed to maintain a sense of meaning, while at the same time observing and trying to understand the environment. In this sense, we arrive at a paradox: While the systems are closed in terms of their self-reference, they are in fact at the same time open with regard to observing the environment. The concept of observation is therefore essential to how systems evolve and change. To actively try to understand and make sense of the environment, the psychological system needs to continually recreate its own self-understanding and sense of meaning. It is on the basis of this self-understanding and sense of meaning that the actual observations are made [22].

For the social systems to influence the psychological systems, and vice versa, the systems need to create noise or distortion in order for the other system to observe and try to make sense of it all. For the psychological system of a person with diabetes to be influenced by the COVID-19 risk communication, it has to see it as relevant to observe–and observing it has to make sense for the self-reproducing meaning-making processes within the psychological system itself. Observing communication from different social systems represents a high level of complexity, as the individual psychological system is exposed to numerous social systems on a daily basis [23].

The basic operation of observing is making a distinction. When a system observes something, it is automatically not observing something else. Rather, it is not observing the rest of the environment. When people with diabetes are observing anything health-related in any given setting, they are automatically not observing something else in relation to the health-related issue and/or the setting. By analyzing how, where and why these distinctions are made, we can analytically sharpen our focus on why people observe what they observe and, perhaps more importantly, why they do not observe what they do not observe [22].

To understand how systems navigate in the complex environment of possible observations, it is necessary to move on from the concept of observations to what Luhmann called expectational structures. By ascribing expected meaning to different contexts or relations, a person will acquire a sense of where he or she will be able to observe meaning in relation to him- or herself or in relation to any given prior observation [23].

3. Results

The analysis produced three main themes. One overarching finding within all three themes concerns the significant differences between how people with T1D and those with T2D experience themselves, their supportive relations and their diabetes in relation to COVID-19 during the lockdown. This overarching finding is present in all of the following themes, which can be seen as different ways of highlighting and interpreting this difference from various angles and perspectives.

In the presentation of the themes, we will not discuss the theoretical elements explicitly, as we will do that in the discussion. All the themes and the elements within the themes did, however, emerge during the analysis and are therefore linked to complexities of communication, social and psychological systems, meaning-making, observational structures and/or expectational structures.

The presented quotes have been translated from Danish to English.

3.1. Altered Self-Observations

Most of the participants had experienced having to change their view of themselves during the lockdown. The reasons for these changes were, however, manifold. The young female interviewee below, with T1D, found it difficult to suddenly be classified as 'vulnerable':

> "I find it difficult to label myself as vulnerable! I normally see myself as a strong person, but I really do need people to take special precautions for me right now. It's hard to relate to myself in all this, when they call us vulnerable or as part of vulnerable groups. They keep saying that Corona isn't dangerous unless you're already ill and vulnerable, so I guess that's me now? Ill and vulnerable?"
> (i2, female, 26y, T1D)

Finding yourself suddenly belonging to a vulnerable group has had a severe impact on all the interviewees with T1D. Many of them had seen themselves as strong independent people who were in control of their disease and their life in general, and most of them had successfully constructed a meaningful self-understanding of themselves based on having T1D. This self-understanding had, in many cases, been severely threatened by this new situation, in which diabetes was suddenly observed by everyone in their environment as a source of a specific vulnerability. Their own observations of the social system observing them as vulnerable basically had forced them to re-create their sense of meaning and, thereby, to change their ways of observing as well as their expectational structures regarding themselves and their environment.

Being in the vulnerable category did not merely entail having an unwanted, ill-fitting label put on you because these people were themselves very worried about the coronavirus–so they also labelled themselves as being in need of special attention. This created an observational paradox: Most of them observed themselves as being at risk, while they at the same time found it difficult to be externally observed as vulnerable. This was especially difficult because internal and external observations will always be strongly interdependent owing to the ongoing creation and re-creation of meaning. They might have no problem identifying with the disease and the increased risks, but the ensuing label was disturbing for these people because many of them had done considerable identity work that would allow them to identify positively with their disease:

"I do want all the love and people being considerate and showing me that they're taking care of me and that they know I need to be treated with caution. But I don't want to be "vulnerable" I want to be taken seriously—but without the label" (i2, female, 26y, T1D)

All of the interviewees with T1D generally identified quite strongly with their disease, and this seems to be the reason why the vulnerable label caused considerable confusion and identification problems. For the interviewees with T2D, the situation was more or less the opposite. Most of them did not identify as someone with a serious disease, and therefore many of them did not identify with belonging to a risk group either.

"I do not feel that I'm a weakened person in any way because I have this diabetes 2. I might be one, but I don't feel it ... Well, I got this feeling of 'oh my god, am I really a sick and weak person.' It stays in me for a short while but then it's out again ... But I do think a little bit about it ... but am I really that bad? Am I really in a risk group? I don't feel that I am, but am I? I'm never sick and I'm not sick more often after getting diabetes." (i19, male, 54y, T2D)

In our study, the interviewees with T2D did not generally observe themselves as people with a serious disease, and therefore they did not normally relate to being vulnerable or belonging to a risk group either. Most of them had no problem discarding the social systems observing them as being at risk, and therefore this label did not affect how they created meaning and self-understanding. They did take the lockdown restrictions seriously, but not because of their diabetes.

"I guess it's because I don't have any real problems with my diabetes. Other people might have, but I must admit, that I'm not the kind of person who takes it that seriously. I take it as it is and it's not like I'm dying because some doctor told me I have this disease. I try to live completely normal." (i11, female, 71y, T2D)

The interviewees with T2D tended to adopt this approach to diabetes and the risk of COVID-19. They understood themselves as not needing to take their diabetes particularly seriously, and they, therefore, did not relate to being in a particular risk group.

Creating meaningful self-understandings and/or changing the expectational structures for self-observations is often closely related to suddenly feeling different from others. Several of the interviewees in our study told us that the precautions they had to take during the lockdown made them feel different:

"When I started wearing the mask people looked at me like: 'Oh my, she's hysterical' Everyone thought I was overreacting when I was the only one in the mall with a mask on. I felt like I needed a big sign on my back reading "I'm the one who is sick" (i18, female, 44y, T1D)

A different and significantly more positive way of observing yourself differently in the lockdown context is as someone who is actually handling it all pretty well. Some of the interviewees reported finding strength and meaning in observing themselves as able to handle difficult personal and societal changes very well:

"This might sound silly, but I actually think that I'm handling all of this pretty well. I think I need to give myself credit for this and to have a bit more self-confidence. I'm pretty damn good at this." (i15, female, 23y, T1D)

This young woman compared herself to her peers in various diabetes peer-to-peer Facebook groups; she had observed a general frustration and fear among the other members. Observing this construction of meaning in a social system of peers had caused her to realize that she stood out from the group, and she then changed her expectational structures and group affiliation accordingly. These changes in her observation of her environment then significantly changed the way she observed herself as competent to cope with COVID-19 fear as well as the lockdown challenges.

3.2. Unclear Risk Perceptions

How the individual with diabetes perceives the combined risk of diabetes and COVID-19 is closely related to how that person perceives diabetes as an integrated part of everyday life and of the above-presented self-understanding.

The pressure of not fully understanding the actual risk was almost unbearable for several of the interviewees with T1D, all of whom tended to identify strongly with their disease. When faced with complex changes in the environment or a general rise in the level of complexity, every system will react by trying to observe identifiable structures and create a sense of meaning. In many cases, the lack of coherent knowledge and firm assessments of the COVID-19 risk for people with diabetes made this creation of meaning difficult.

"I can't not know anything. The pressure of not knowing the risk is greater than the fear of contracting it some days. You just get a fXXX-it-all kind of feeling and end up giving up on yourself and your health–and I think that's the worst that can happen. That's the kind of pressure you get from not understanding anything." (i6, female, 27y, T1D)

If something is perceived as very risky, thereby threatening the way you normally create a meaningful sense of yourself, it can significantly change self-understandings and structures of meaningful observations.

This suddenly emerged new complexity was also evident on the level of everyday diabetes management. The interviewees (again, especially those with T1D) could not seem to regulate their blood glucose in the way they had before the lockdown–due to changes in everyday routines. This caused more hands-on worries:

"Yes, of course, I'm a little worried. It's been difficult to regulate myself. I have a feeling that the more unstable your blood glucose is the worse your immune system is–and that makes me think that the virus might find an easier way in with someone who is poorly regulated. So that's the reason I've been more scared or nervous–I've been poorly regulated" (i7, male, 20y, T1D)

Thus, the interviewees did not only have to come to terms with a virus that puts them at a higher risk than the rest of the population. They also had to try to navigate their daily disease regulation through unknown territory, while trying to constantly reduce the complexity of the whole situation affecting how they saw themselves, how they observed their disease and how they made sense of their whole environment, as defined by societal systems now difficult to observe meaningfully.

This apparent state of confusion among the interviewees with T1D regarding risk seemed to be further enforced by the frustrations shared in Facebook groups:

" ... and I hear from others that there are several with T1D who get sick and contact their doctor because they're also poorly regulated and therefore in the official risk group. But then the doctor refuses them a corona test and that's just not OK ... They are just being told that they're young and fit and therefore not at risk. We're just not being taken seriously" (i2, female, 26y, T1D)

This shared identity of belonging to a risk group seemed to be both good and bad in relation to risk perceptions and being able to positively relate to diabetes. There was a distinct feeling of being ignored as a risk group or even left out of the closed-down society, which contributed to negative expectation structures that seemed to then turn into negative views on diabetes identification and negative self-understandings. On the other hand, it did seem to provide common ground in a time of solitude and potential loneliness.

> "It's just really nice to write with and talk to my girlfriends who also have diabetes, because they feel the exact same way. I mean, everyone else understands absolutely nothing about how we feel right now. No one informs us about anything, and we're all gaining weight 'cause we've eaten too much cake like the rest of Denmark. It's just really nice to feel understood" (i2, female, 26y, T1D)

The situation was, once again, quite different for most of the interviewees with T2D. In most cases, they did not observe themselves as being at an increased risk, and several of them did not acknowledge any noteworthy connection between T2D and COVID-19:

> "From my perspective, I just can't see what the two things have got to do with each other ... I mean, I do understand that if you're a diabetes patient who eats four bags of crisps every day and weigh three times what you're supposed to and never leaves the house and everything else you can do wrong, then 'yes' maybe your world will fall apart! But for myself, I mean the last thing I ever think about is the fact that I somehow got this diabetes a couple of years ago–it has absolutely no influence on my daily life" (i3, male, 58y, T2D)

Even if most of the interviewees with T2D ignored diabetes in the context of COVID-19, there was a tendency for those with T2D who also had other reasons for being at increased risk to take it seriously. In these cases, however, the main reason for taking the risk seriously was rarely the diabetes:

> "I definitely worry. It doesn't sound pleasant at all–even the mild cases. I mean, with my age, 73, and hypertension and now diabetes, I'm in the risk group. So yeah, I'm a bit worried." (i8, male, 73y, T2D)

The general tendency among the interviewees with T2D, however, was to not take it particularly seriously. Many of them told us that their relatives worried much more than they themselves did:

> "Our kids believe that I need to be extra careful. They tell me on a regular basis that now I'm in the danger group or whatever it's called? Risk group. They worry about me more than I do myself" (i10, male, 69y, T2D)

Even when the media placed people with T2D firmly in the risk group and the people close to them tended to express worries about the relation between T2D and COVID-19, the interviewees with T2D tended to significantly downplay the severity of the situation, basically observing themselves as not belonging to a particular risk group. The major reason for this seemed to be that they did not identify with their disease in the first place, and therefore their self-understanding was not strongly connected with them having a chronic disease.

3.3. Changes in Relational Everyday Life

How people deal with abrupt changes and/or increased complexity is deeply dependent on their individual ability to relate to communication in social systems. What happens when the basic aspects of relational everyday life change from one day to the next? How do people with diabetes then cope with general meaning-making and basic self-observation?

> "It really is kinda unsettling when everyone around you suddenly goes: You really have to take special care now'" (i16, male, 68y, T2D)

Even though many of the interviewees with T2D had done their best to ignore and dismiss their heightened risk, they were also the ones who were most frustrated about the discrepancies between their view of themselves as healthy and not at risk and how the environment tended to view them as being at risk. They were basically observing the environment's observations of them as out of tune with the meaning they had created through their own observation. This created a sense of meaning-crisis in several of them, and they seemed especially vulnerable to these changes in relational and public perceptions. Immediately before her interview, one female interviewee with T2D had just spoken with her son about this, and the conversation had changed the way she observed herself in relation to risk:

"This happened just an hour ago. I had to concede, and I just thought: 'Oh my god, he's right.' I mean, we have to take care of ourselves in this vulnerable situation we're in. I am in the risk group. So, it's actually fantastic that our son thinks about it. It warms my heart that we have children who think like that" (i4, female, 71y, T2D)

Being able to meaningfully observe the worries and underlying affection of close relations seemed to create the basis for changes in both self-observations and risk perceptions.

Observing yourself being observed by the media (and society as a whole) as vulnerable or at risk had left a strong impression on many of the interviewees:

"This is just so difficult to change. If you have a population that possesses a piece of knowledge that they believe is true, then it really takes so much to change that in any way ... So, no matter how much empathy or love you have for a given disease population, you will, in some way be affected by what the rest of society thinks about this disease ... And I guess we're just really unlucky in that respect" (i2, female, 26y, T1D)

This relationship with the rest of society was a burden, especially for the interviewees with T1D. They often felt a general lack of knowledge about what diabetes is and does. Further, they felt that this lack of knowledge strongly affected how they were observed by society and, consequently, how they ultimately observed themselves. This situation of observing close and distant relations in the complex network of social systems was one of reducing complexity by deciding or knowing where to observe, so as to continually be able to create a sense of meaning that corresponds with the sense of self they were simultaneously creating through their perpetual self-observations.

In order to create this sense of meaning, many of them had placed a great deal of value on their peer relationships:

"Yesterday I was chatting with some friends who also have diabetes and I just thought that I'm so lucky to have them, because, just imagine all the people with diabetes who don't know anyone else with diabetes ... They're just extra alone right now and that must be really hard." (i2, female, 26y, T1D)

Creating communicative meaning by seeking out well-known and well-established relationships in times of crisis and insecurity was a way for especially the younger interviewees with T1D to reduce communicative complexity. This appeared to have created a situation with stronger internal identification and feelings of belonging in these groups, while also weakening external identification and creating a feeling of being left out or forgotten.

The lockdown and the perception of being at risk had also affected routines and the everyday understanding of roles in close relationships for the interviewees with T1D:

"I do understand that it's frustrating to live with a diabetic right now, because there's really not a lot I can help him with. Normally we're like an old married couple with certain things each of us does more than the other. We've divided the chores and that's just completely [obscenity] up right now. I don't do any shopping, I don't see any friends, I can't go to work and since I'm working freelance, I make less money. I can't go to uni to meet with my study partner 'cause I just think it's too dangerous." (i6, female, 27y, T1D)

This feeling of having your everyday life turned upside down was something all the interviewees with T1D described. For many of them, this was linked to the frustration of not having received any definite risk assessments from the authorities, which had turned into a negative spiral of regrets, doubts and fears that life as they knew it was falling apart. They felt this put great pressure on their important relations and, as the young woman in the above quote clearly described, these doubts had caused them to constantly observe how they were being observed by their loved ones. This made them feel like a burden, which again fed the negative spiral with more and more negative self-observation.

4. Discussion

In our study, we have identified three highly interdependent themes of relevance to how people with diabetes navigate living with the disease during the COVID-19 lockdown and how this affects their sense of self and of being a part of society. These themes are all strongly related to structures of observation, or to be more precise, structures of observing observation [22]. The interviewees all mentioned how they had tried to make sense of communication from health authorities, mass media, peer groups and family members, while at the same time trying to make sense of the way these communicative sources were observing people with diabetes as belonging to risk groups and/or being vulnerable. This mutual process of observations is one that is ongoing at any time and in any society, but as the results of our study show, the challenges of having to create and re-create meaning internally (psychological systems) and externally (social systems) can be especially demanding at times when very basic elements of meaning, and of how we are able to observe each other, are changing dramatically. These findings add significant new knowledge about when and how to apply theories about communicative social systems and maybe most of all highlights the potentials in applying the concept of observation to how vulnerable groups react to societal developments [21–23].

These experiences of living in a society that is locked down illuminate the mechanisms of psychological systems trying to create meaning based on observations of social systems. The fact that normal everyday interactions with friends, colleagues, family members and the healthcare professionals suddenly are extremely limited, creates a new and unknown situation, where this meaning-making is severely challenged–causing insecurities and the personal need for complexity-reducing health- and risk information.

When traditional structures for observing are no longer operational, the meaning-making process of creating expectational patterns as a way of reducing complexities in daily life is becoming a daily struggle for the people with diabetes, not knowing what the future brings or how to create meaning based on the observed communication. Most significantly this lack of meaning regards being placed in a group defined and observed as vulnerable or at risk.

In the study, we saw significant differences between persons with T1D and those with T2D in regards how they experience themselves, their supportive relations and their diabetes in relation to COVID-19 during the lockdown, but we have also shown the general (both T1D and T2D) ambivalence associated with being observed as belonging to a vulnerable group. Whereas the interviewees with diabetes created meaning out of the situation in a multitude of ways, they all responded to being labelled as persons at risk. Having to suddenly observe themselves being observed in a very different way than before had forced them to adapt to this observation by adjusting how they observe themselves. This significant finding has not been reported in other studies so far and is, therefore, a valuable contribution to the existing research in COVID-19 and diabetes.

All 20 participants had different disease trajectories that could potentially impact their reaction to COVID-19 and the lockdown. Apart from the significant differences between the two main types of diabetes, it could be assumed that an aspect like years since diagnosis would play a role in how people with diabetes manage their disease in a time of change and insecurities. We did, however, not find strong indications of that.

Coming from a different theoretical angle than Luhmann did, philosopher Ian Hacking argued that such labels may have 'looping effects' in so far as people change their behaviours and self-understandings in response to the categories in which they are placed [27]. In the present case, the categorization of people with diabetes as a vulnerable group appears to have had consequences for their behaviours, in terms of diabetes management and social isolation, as well as for the extent to which they identified with their disease. The effects become 'looping' because these changes reflect back on their vulnerability by transforming the very conditions that made them vulnerable in the first place. Hacking's theory supports our findings, and further research might benefit from an analytical strategy that combines Hacking's and Luhmann's respective approaches.

Our findings have important implications for health communication in general and risk communication in particular. They show that people may respond in a multitude of ways to communications issued by health authorities and that their meaning-making is shaped by, and shapes, their self-observations, risk perceptions and close relational environments. This may call for more targeted communication strategies as well as increased use of peer support that can help people create meaning within their own environments. This is relevant to both micro-level communications with the diabetes clinic or the GP and macro-level communication on the societal level. In fact, this split-level communication strategy is especially pertinent to people with diabetes, as they are exposed to the massive amount of risk communication dealing with COVID-19 risk on a more general level and need diabetes-specific clarifications from specialist sources.

As we wished to study people with diabetes and their immediate reactions to the COVID-19 lockdown, we had to plan and perform the interview study within a very short period of time. This meant that things were done more quickly than would normally be the case. We focused on eliminating rushed mistakes by keeping everything as simple as possible, though without compromising our research standards. We decided to keep the interview guide very basic and open, as we did not have sufficient background literature to target specific pre-themes or pre-understandings. If we had conducted the study two or four months later, we would have been able to make it more specific because there would have been more time for preparation and more knowledge about COVID-19 and societal reactions in general. On the whole, however, we are satisfied with the richness and depth of the data.

The external validity or transferability of the study at hand is somewhat difficult to assess given the novelty of the societal situation studied. To what extent our findings can be applied beyond the contexts in which the studies were done, and how a more optimal sample might be constructed in order to make the findings more transferable is indeed rather difficult to say. Regarding recruitment, we decided to include people with different types of diabetes in the same study. We did this because we did not know what we would find. Although the two types of diabetes (and the daily tasks of managing the diseases) are significantly different, we did not know whether this would transfer to reactions to the risks associated with COVID-19. Another reason for including both types was that the general risk information issued by health authorities in the early stages of the pandemic was generally not specific to diabetes type. In retrospect, the differences between the two groups helped highlight the main findings of our study. The reactions to COVID-19 and the lockdown, however, were so different that future studies could benefit from a more specific focus on the characteristics of the two main types of diabetes. This would give us a more in-depth understanding of some of the elements that were specific to T1D and T2D.

Another aspect of the transferability is the choice of theory. Working with the theory in our study makes it clear that the actual choice of theory strongly affects the degree of transferability. This is very much in line with Rasmussen's [24] description of how, within radical hermeneutics, there is a constant constructivist awareness of the relationship between theory and the construction of the environment observed. Employing a theory that focuses on complexity makes the findings more flexible and adaptable to different contexts and settings, which is especially relevant in a situation, where we are studying a novel and previously unknown societal phenomenon.

Author Contributions: Conceptualization, D.G., J.M. and M.H.R.; Formal analysis, D.G.; Investigation, D.G. and J.M.; Methodology, D.G., J.M. and M.H.R.; Writing—original draft, D.G.; Writing—review & editing, J.M. and M.H.R. All authors have read and agreed to the published version of the manuscript.

Funding: This research received no external funding.

Conflicts of Interest: The authors declare no conflict of interest.

References

1. Cucinotta, D.; Vanelli, M. WHO Declares COVID-19 a Pandemic. *Acta Biomed.* **2020**, *91*, 157–160.
2. Rothan, H.A.; Byrareddy, S.N. The epidemiology and pathogenesis of coronavirus disease (COVID-19) outbreak. *J. Autoimmun.* **2020**, 102433. [CrossRef] [PubMed]
3. Diabetesforeningen [Danish Diabetes Association]. Sundhedsstyrrelsen: Alle med Diabetes er i Risikogruppen [Danish Health Authority: All People with Diabetes Are at Risk]. 2020. Available online: https://diabetes.dk/aktuelt/nyheder/nyhedsarkiv/2020/sundhedsstyrelsen-alle-med-diabetes-er-i-risikogruppen.aspx (accessed on 18 July 2020).
4. Kumar, A.; Arora, A.; Sharma, P.; Anikhindi, S.A.; Bansal, N.; Singla, V.; Khare, S.; Srivastava, A. Is diabetes mellitus associated with mortality and severity of COVID-19? A meta-analysis. *Diabetes Metab. Syndr.* **2020**, *14*, 535–545. [CrossRef] [PubMed]
5. Wang, X.; Wang, S.; Sun, L.; Qin, G. Prevalence of diabetes mellitus in 2019 novel coronavirus: A meta-analysis. *Diabetes Res. Clin. Pract.* **2020**, *164*, 108200. [CrossRef] [PubMed]
6. Fadini, G.P.; Morieri, M.L.; Longato, E.; Avogaro, A. Prevalence and impact of diabetes among people infected with SARS-CoV-2. *J. Endocrinol. Investig.* **2020**, *43*, 867–869. [CrossRef] [PubMed]
7. Huang, I.; Lim, M.A.; Pranata, R. Diabetes mellitus is associated with increased mortality and severity of disease in COVID-19 pneumonia—A systematic review, meta-analysis, and meta-regression. *Diabetes Metab. Syndr.* **2020**, *14*, 395–403. [CrossRef]
8. Cuschieri, S.; Grech, S. COVID-19 and diabetes: The why, the what and the how. *J. Diabetes Complicat.* **2020**, *34*, 107637. [CrossRef]
9. Gupta, R.; Hussain, A.; Misra, A. Diabetes and COVID-19: Evidence, current status and unanswered research questions. *Eur. J. Clin. Nutr.* **2020**, *74*, 864–870. [CrossRef]
10. Katulanda, P.; Dissanayake, H.A.; Ranathunga, I.; Ratnasamy, V.; Wijewickrama, P.S.A.; Yogendranathan, N.; Gamage, K.K.K.; de Silva, N.L.; Sumanatilleke, M.; Somasundaram, N.P.; et al. Prevention and management of COVID-19 among patients with diabetes: An appraisal of the literature. *Diabetologia* **2020**, 1–13. [CrossRef]
11. Singh, A.K.; Gupta, R.; Ghosh, A.; Misra, A. Diabetes in COVID-19: Prevalence, pathophysiology, prognosis and practical considerations. *Diabetes Metab. Syndr.* **2020**, *14*, 303–310. [CrossRef]
12. Targher, G.; Mantovani, A.; Wang, X.B.; Yan, H.D.; Sun, Q.F.; Pan, K.H.; Byrne, C.D.; Zheng, K.I.; Chen, Y.P.; Eslam, M.; et al. Patients with diabetes are at higher risk for severe illness from COVID-19. *Diabetes Metab.* **2020**. [CrossRef]
13. Guo, W.; Li, M.; Dong, Y.; Zhou, H.; Zhang, Z.; Tian, C.; Qin, R.; Wang, H.; Shen, Y.; Du, K.; et al. Diabetes is a risk factor for the progression and prognosis of COVID-19. *Diabetes Metab. Res. Rev.* **2020**, e3319. [CrossRef] [PubMed]
14. Zhou, F.; Yu, T.; Du, R.; Fan, G.; Liu, Y.; Liu, Z.; Xiang, J.; Wang, Y.; Song, B.; Gu, X.; et al. Clinical course and risk factors for mortality of adult inpatients with COVID-19 in Wuhan, China: A retrospective cohort study. *Lancet* **2020**, *395*, 1054–1062. [CrossRef]
15. Singh, A.K.; Singh, R. Does poor glucose control increase the severity and mortality in patients with diabetes and COVID-19? *Diabetes Metab. Syndr.* **2020**, *14*, 725–727. [CrossRef] [PubMed]
16. Zhang, Y.; Li, H.; Zhang, J.; Cao, Y.; Zhao, X.; Yu, N.; Gao, Y.; Ma, J.; Zhang, H.; Zhang, J.; et al. The clinical characteristics and outcomes of diabetes mellitus and secondary hyperglycaemia patients with coronavirus disease 2019: A single-center, retrospective, observational study in Wuhan. *Diabetes Obes. Metab.* **2020**, *22*, 1443–1454. [CrossRef]
17. Adhikari, S.P.; Meng, S.; Wu, Y.J.; Mao, Y.P.; Ye, R.X.; Wang, Q.Z.; Sun, C.; Sylvia, S.; Rozelle, S.; Raat, H.; et al. Epidemiology, causes, clinical manifestation and diagnosis, prevention and control of coronavirus disease (COVID-19) during the early outbreak period: A scoping review. *Infect. Dis. Poverty* **2020**, *9*, 29. [CrossRef]
18. Mainous, A.G. A towering babel of risk information in the COVID-19 pandemic: Trust and credibility in risk perception and positive public health behaviors. *Fam. Med.* **2020**, *52*, 317–319. [CrossRef] [PubMed]
19. Statsministeriet [The Prime Minister's Office]. Pressemøde om COVID-19 den 11. marts 2020 [Press Conference on COVID-19 the 11th of March 2020]. Available online: https://www.stm.dk/_p_14920.html (accessed on 22 July 2020).
20. Rudnick, A. Social, psychological and philosophical reflections on pandemics and beyond. *Societies* **2020**, *10*, 42. [CrossRef]

21. Luhmann, N. *Essays on Self-Reference*; Columbia University Press: New York, NY, USA, 1990.
22. Luhmann, N. *Social Systems*; Stanford University Press: Palo Alto, CA, USA, 1995.
23. Luhmann, N. *Theories of Distinction*; Stanford University Press: Palo Alto, CA, USA, 2002.
24. Rasmussen, J. Textual interpretation and complexity—Radical hermeneutics. *Nord. Pedagog.* **2004**, *24*, 177–194.
25. Carlsson, S. Environmental (Lifestyle) risk factors for LADA. *Curr. Diabetes Rev.* **2019**, *15*, 178. [CrossRef]
26. Malterud, K.; Siersma, V.D.; Guassora, A.D. Sample size in qualitative interview studies: Guided by information power. *Qual. Health Res.* **2016**, *26*, 1753–1760. [CrossRef] [PubMed]
27. Hacking, I. The looping effects of human kinds. In *Causal Cognition. A Multidisciplinary Debate*; Sperber, D., Ed.; Clarendon Press: Oxford, UK, 1995; pp. 351–394.

© 2020 by the authors. Licensee MDPI, Basel, Switzerland. This article is an open access article distributed under the terms and conditions of the Creative Commons Attribution (CC BY) license (http://creativecommons.org/licenses/by/4.0/).

Concept Paper

Online Learning and Emergency Remote Teaching: Opportunities and Challenges in Emergency Situations

Fernando Ferri, Patrizia Grifoni and Tiziana Guzzo *

Institute for Research on Population and Social Policies, National Research Council, 00185 Rome, Italy; fernando.ferri@irpps.cnr.it (F.F.); patrizia.grifoni@irpps.cnr.it (P.G.)
* Correspondence: tiziana.guzzo@irpps.cnr.it

Received: 28 August 2020; Accepted: 5 November 2020; Published: 13 November 2020

Abstract: The aim of the study is to analyse the opportunities and challenges of emergency remote teaching based on experiences of the COVID-19 emergency. A qualitative research method was undertaken in two steps. In the first step, a thematic analysis of an online discussion forum with international experts from different sectors and countries was carried out. In the second step (an Italian case study), both the data and the statements of opinion leaders from secondary online sources, including web articles, statistical data and legislation, were analysed. The results reveal several technological, pedagogical and social challenges. The technological challenges are mainly related to the unreliability of Internet connections and many students' lack of necessary electronic devices. The pedagogical challenges are principally associated with teachers' and learners' lack of digital skills, the lack of structured content versus the abundance of online resources, learners' lack of interactivity and motivation and teachers' lack of social and cognitive presence (the ability to construct meaning through sustained communication within a community of inquiry). The social challenges are mainly related to the lack of human interaction between teachers and students as well as among the latter, the lack of physical spaces at home to receive lessons and the lack of support of parents who are frequently working remotely in the same spaces. Based on the lessons learned from this worldwide emergency, challenges and proposals for action to face these same challenges, which should be and sometimes have been implemented, are provided.

Keywords: online learning; emergency remote teaching; technological challenges; pedagogical challenges; social challenges

1. Introduction

The coronavirus (COVID-19) was declared a global pandemic on 12 March 2020 and social distancing was adopted in many places to contain the problem. Indeed, numerous countries around the world decided to close schools nationwide to prevent or contain the spread of the virus, significantly affecting the learning of millions of children and adolescents. COVID-19 has highlighted the problem of the management of school lessons and learning processes worldwide, among issues. Technology can certainly be of support in this regard.

Ministries of education in different countries have recommended or made it mandatory to implement online learning at all school levels in various countries. This decision has also been supported by UNESCO [1], which has declared that online learning can help stop the spread of the virus by avoiding direct interactions between people. UNESCO [2] has additionally provided a list of free educational platforms and resources that can be used for online learning according to the needs of each educational institution, providing social care and interaction during school closures.

· Online learning can be defined as instruction delivered on a digital device that is intended to support learning [3]. In the literature, several advantages of online learning have been highlighted: studying from anywhere, at any time; possibility of saving significant amounts of money; no commuting on crowded buses or local trains; flexibility to choose; and saving time [4–6]. Online learning is thus becoming more and more important for education during the time of the worldwide health emergency, offering the opportunity to remain in touch, even if remotely, with classmates and teachers and to follow lessons. However, many challenges have been observed in different countries. The most evident and widely discussed by experts and policymakers is that socially disadvantaged groups face difficulties in meeting the basic conditions required by online learning [7]. The next section introduces previous studies on online learning in emergency situations. Lockdowns and the subsequent closure of educational institutions seem to have amplified the gap between rich and poor people, not just between the Global North and the Global South, but also within countries. School closures could have a negative impact on learners from lower socioeconomic backgrounds, widening the gap with their more advantaged peers [8]. Indeed, on the one side, there is the main objective of safeguarding health, while on the other side the aforementioned problems are emerging.

The adoption of online learning in a situation of emergency represents a need, but it has also stimulated experts, policymakers, citizens, teachers and learners to search for new solutions. This is producing a shift from the concept of online learning to emergency remote teaching, which represents "a temporary shift of instructional delivery to an alternate delivery mode due to crisis circumstances" [9]. As stated by UNESCO Director-General Audrey Azoulay: *"We are entering uncharted territory and working with countries to find hi-tech, low-tech and no-tech solutions to assure the continuity of learning"*. For this reason, new challenges and opportunities at a social and technological level may emerge. It is an experience that enables us to reflect on the different approaches and lessons learned in different countries and additionally provides an opportunity to find new solutions. In fact, greater reflection on and study of social challenges related to the current pandemic and more generally to global crises are necessary [10].

There are several studies on online learning during emergencies. Besides confirming and reinforcing the challenges identified by previous research, our study provides a framework on the opportunities, challenges and lessons learned in different countries during the COVID-19 emergency, with a special focus on Italy. Although previous investigations have offered some paths to follow, they do not provide specific actions deriving from lessons learned. Our study aims to contribute to filling this gap. Indeed, starting from the previous works, and enriched by an online discussion forum and data from secondary sources regarding Italy, we extract challenges and proposals for action to face these same challenges, that different actors (policymakers, professors, etc.) should implement to face the ongoing and emerging challenges. United Nations [11], in fact, in order to mitigate the negative consequences of the COVID-19 on education, are encouraging governments and stakeholders to take actions and to accelerate changes in modes of delivering quality education, leaving no one behind. According to [11], inclusive changes in education delivery through education investment and reforms at the governance level are necessary. This pandemic can be an opportunity and an exercise for emergency remote teaching to evaluate emerged challenges during emergencies and develop a coherent online education strategy for any other emergencies or natural disasters that can potentially happen in the future. This is also underlined by UNESCO: [12] "Education systems around the world are facing an unprecedented challenge in the wake of massive school closures mandated as part of public health efforts to contain the spread of COVID-19. Governmental agencies are working with international organizations, private sector partners and civil society to deliver education remotely through a mix of technologies in order to ensure continuity of curriculum-based study and learning for all". Furthermore, one of the aims of The Global Education 2030 Agenda of UNESCO [13] is the quality education which aims to "ensure inclusive and equitable quality education and promote lifelong learning opportunities for all". In this context, more attention is necessary on how technology and learning can be integrated effectively, including the vital role of teachers, and the students' needs. Therefore, it is very important to analyse challenges related to emergency remote teaching and indicate proposals for action to face these challenges; these proposals are addressed to government, decision-makers and stakeholders in order to guaranty quality in education. The purpose of the manuscript is to reflect on the summary of opinions by experts coming from an online discussion forum and the data analysis of the Italian case study, and which served to substantiate proposals for action as crucial to meeting the new challenges listed.

For this reason, a qualitative research method was undertaken in two steps. In the first step, a thematic analysis of an online discussion forum was carried out. This forum was organised in the framework of the HubIT project [14] and involved experts concerning the challenges faced during the current crisis. In the second step, we carried out an analysis of secondary online sources, like web articles, statistical data and legislation, to analyse our Italian case study.

The paper is organised as follows. Section 2 provides an analysis of the literature on criticalities and the challenges of emergency remote teaching. In Section 3, the methodology is described. Section 4 discusses the challenges that emergency remote teaching is presenting, identified during the forum discussion. Section 5 describes the Italian case study through an analysis of secondary sources. Finally, Section 6 concludes the paper.

2. Related Works

Despite the crisis produced by COVID-19, online learning has enabled many people to continue teaching and learning without interruption. The pandemic crisis is the reason for the widest experimentation in online education globally. However, a systematic approach to understanding the pros and cons of online learning and for investing, planning and delivering it is necessary, given its broad implementation and expansion [15].

During the school closures, existing inequalities connected to different socioeconomic situations have increased mainly due to the following reasons: (i) lack of resources, including access to educational technologies and the Internet; and (ii) lack of physical spaces to carry out home-based learning among families from poorer backgrounds, who lack the basic skills to support their children, especially regarding secondary education [16–19]. There is some evidence that school closures can produce significant losses in educational achievement, in particular for disadvantaged students [7].

In the Netherlands, these factors have resulted in a large gap in how children have been learning during this emergency period [17]. In developing countries like Ghana, in which the majority of students do not have access to the Internet and adequate learning environments, such discrepancies are even more apparent [20]. Similar challenges have also been faced also in Malaysia. To overcome these difficulties, Yusuf [21] suggests that institutions should provide more adequate e-learning platforms to increase access to the Internet and develop an interactive learning approach. Moreover, it is necessary to provide workshops or training for teachers and students to improve their technological and pedagogical competencies in online learning. The question of inclusion is central when we consider emergency remote teaching. Inclusion may have different characteristics across countries. For example, in South Africa inclusion is connected with processes of de-colonisation. Indeed, according to Omodan [22], there is the need to decolonise rural universities in South Africa, to be able to respond to every unforeseen emergency, as an outcome of coloniality.

The advantages and limitations of using online learning in medical and dental institutes in Pakistan have been analysed [23]. This study found that online learning was a flexible and effective source that allows students to become self-directed learners, although disadvantages related to the inability to teach and learn practical and clinical work were also highlighted. Another criticality was represented by the lack of immediate feedback for students. In response, the authors recommended training faculty and developing lesson plans with reduced cognitive load and increased interactivities. According to Verawardina et al. [24], it is necessary to implement clear steps in applying online learning, such as preparing facilities, training with current technology, providing guidelines for teachers and students, offering interactive multimedia materials in line with the current curriculum and ensuring an evaluation system with a question bank.

It is important to view learning not as a process of information transfer but as a social and cognitive process [9]. In the planning of online learning, it is necessary to model not only the content but also the different interactions that occur in this process. In fact, Bernard et al. [25] have found that interactions increase learning outcomes.

This pandemic may accelerate some changes in educational models based on the pros and cons of the technology used for learning purposes. Thomas and Rogers [18], starting from their experiences of online learning during the pandemic emergency, have observed that school-provided IT systems are frequently too expensive, cumbersome and quickly go out of date. They suggest moving to personal devices integrated into schools. Moreover, they recommend that policymakers incentivise and encourage companies to produce engaging and powerful educational games and learning environments. To gamify education will encourage children's engagement and curiosity. Eder [26] additionally suggests using television or radio for online learning in order to reach learners who lack access to the Internet, although this requires time to plan and programme content. Nevertheless, it is worth noting that different media like radio and television were also used in 2014 during the Ebola crisis [27]. Furthermore, during the current crisis, some countries have used different modalities for online learning to avoid the problem of the digital divide. New Zealand, for example, has adopted a combined approach, using two television channels to deliver educational content, integrated with an Internet delivery and a hard-copy curriculum resource. In Queensland (Australia), due to poor Internet connectivity television has been used to engage parents as well so that they can assist their children in learning. In Portugal, hard-copy teaching resources have been promptly delivered to children's homes thanks to a partnership between schools and post office services [28].

Table 1 summarises some key obstacles to the effective use of online learning identified in the literature.

Table 1. Open challenges of online learning.

OPEN CHALLENGES		REFERENCES
TECHNOLOGICAL CHALLENGES	Access to infrastructure such as technological devices and an Internet connection.	[8,16–18,20–22,28]
PEDAGOGICAL CHALLENGES	Teachers' lack of skills in using technology. Need for training and guidelines for teachers and students.	[21–24]
	Need for teaching materials in the form of interactive multimedia (images, animations, educational games) to engage and maintain students' motivation.	[18,23,24]
	Lack of student feedback and evaluation system.	[18,23,24]
SOCIAL CHALLENGES	Lack of suitable home learning environment to study and parents' support.	[8,16–19]

3. Materials and Methods

In order to analyse experiences, opportunities, open challenges and lessons learned regarding online learning during the COVID-19 emergency, a qualitative method was used based on a two-step process. The first step consisted of an online discussion forum. This forum was organised to include researchers, professors and enterprises mainly from European countries and from Lebanon with expertise in information and communications technology (ICT), social science and education. The discussion enabled the participants to discuss and compare their experiences, primarily related to the COVID-19 crisis. We collected their opinions and experiences in a narrative way. The results from the forum represented the basis for the second step, in which starting from the main issues that emerged from the online discussion, we undertook an analysis of secondary online sources (web articles, statistics, legislation) about Italy. Indeed, Italy was the first European country to undergo a long period of lockdown. Therefore, we decided as an Italian partner of the HubIT project to follow the debate among the opinion leaders in Italy and also to consider data from the Istituto Nazionale di Statistica (Italian National Institute of Statistics, ISTAT) regarding distance learning. The two previous steps enabled us to gain a complete picture of different statements from diverse perspectives and experiences. In these ways, we analysed and understood the challenges and opportunities as well the perceived need to accelerate innovation in online learning, considering pedagogical, social and technological points of view.

In the first step, we opted to conduct a discussion using an online discussion forum, as it represents a feasible alternative to traditional face-to-face focus groups [29]. In this lockdown period, online forums allowed a wide diversity of participants to join in the discussions and interact with each other from different geographical areas, exchanging their experiences and opinions without requiring long-distance travel. Furthermore, the public accessibility of the analysis material allows other researchers to retrace the analysis process, adding greater transparency than in other qualitative methods.

In particular, we organised the discussion forum titled "Distance learning and emergency remote teaching: opportunities and criticalities at the time of the worldwide health emergency-SPEAK OUT!" in the framework of the H2020 project "The HUB for boosting the responsibility and inclusiveness of ICT-enabled research and innovation through constructive interactions with SSH research–HubIT" [30].

The sample of the study was selected by using a purposive sample, as is used when the opinion of experts concerning a particular topic of interest is considered necessary. This sampling is also called judgmental sample and it is a type of nonprobability sample; indeed, non-random criteria are used to decide units that should be included in the sample. In this study, we used, in particular, an expert sampling, and although it has the advantage to take opinions or assessment of people with a high degree of knowledge about the topics of interest, it has also some criticisms. If this sample is not representative of the entire population, then the findings are still not generalizable to the overall population at large, even if we invited people representative of the quadruple helix, in order to include different points of view.

Experts were thus chosen based on having characteristics relevant to the study and on being informative. In particular, experts were selected among the project partners of the HubIT project (which involves researchers, professors and industry representatives) and stakeholders identified among professors and experts in ICT, social science and education. The discussion enabled the participants to discuss and compare their experiences mainly connected with the COVID-19 crisis. When invited, participants were informed about the aims of the online forum and were asked to contribute to it and they voluntarily accepted.

To participate in the online forum, experts were required to register on METROPOLIS [14], the platform of the European project HubIT. The online discussion forum took place on 21 May 2020. Fifteen people from Portugal, the United Kingdom (UK) Italy, Estonia, Slovakia, Lebanon and Hungary attended it and a total of 162 comments were collected. This discussion forum is available at [30].

A semi-structured interview guide was used with questions pertaining to the opportunities and limitations of emergency online learning and related challenges such as the need for personal devices and an Internet connection, inclusion and accessibility problems, building a sense of community between learners and teachers, the use of interactive and engaging lessons and the potential use of emerging technologies. The forum was mediated by three moderators, who asked questions and addressed the discussion. Questions were addressed during the discussion forum; any new question was asked when the discussion on a previous question appeared to be closed as no one among the participants provided any new contribution. The method of analysis chosen was thematic analysis. Generally, thematic analysis is the most widely used qualitative approach to analyse data or information. It is used for "identifying, analysing, and reporting patterns (themes) within the data" [31]. The information collected was analysed and categorised, revealing three major themes: (a) technological challenges, (b) pedagogical challenges and (c) social challenges. Qualitative research relies on unstructured and non-numerical data. In this article, we present the results of the first step through selected quotations that are most representative of the research findings.

In the second step, in order to extend and reinforce the online discussion, we analysed a concrete case study of a country of the project partners. Italy was chosen because, in the first phase of the pandemic's spread, it had the second-most COVID-19 cases after China, producing a long period of lockdown, with a strong impact on education. With this work we aim to provide a picture of the implications of distance learning during this emergency in this country, while also analysing statistics regarding distance learning and the digital divide from ISTAT [32], the policies implemented and the statements of opinion leaders in the sector. Experts were selected after an online review of the most significant sources in academia, press and scientific research. The most accredited names on the topic were considered the most relevant for the study. The results of the second step of the study are presented both through an analysis of the different statements of opinion leaders and through statistics from ISTAT [32] about the digital divide in Italy, to provide a picture of the Italian case study.

4. Results of the Forum Discussion: Open Challenges of Emergency Remote Teaching

Emergency remote teaching has given a significant boost to online learning, opening up new opportunities and reflections for the educational system. According to the discussion carried out within the forum, the COVID-19 crisis experience is presenting different challenges that should be addressed to develop new methodologies and pedagogical approaches, infrastructure and platforms specifically designed for online teaching. These new methodologies need to be developed in an interdisciplinary and holistic perspective that (following the responsible research and innovation approach) will anticipate and assess potential implications and social expectations [33].

Indeed, the COVID-19 emergency has made clear that technologies alone do not represent a panacea. The long-term inequality gaps between students in different situations in education systems have frequently been highlighted during the COVID-19 pandemic period. Students and teachers have faced different obstacles in remote teaching due to the existing limitations related to technological, pedagogical and social challenges, which will be analysed in the following sections.

4.1. Technological Challenges

Technological challenges are primarily related to a lack of Internet connectivity and electronic devices. This problem may increase inequalities through uneven access to the technology needed by students and teachers. Indeed, not all learners have access to the necessary technologies to take advantage of online education such as a fast Internet connection and a powerful computer. During the forum discussion a very frequent situation in families with children was described:

> *Just think of families where there is more than one child in school with no or one computer. This means that in parallel only one child can take part in a digital online education course.*

These issues especially affect many disadvantaged families, but also middle-class families with multiple children, or parents who are engaged in smart working. Numerous initiatives have been organised in some European Union (EU) member states to overcome this situation:

> *In Hungary, we have initiated the collection of offers of ICT companies. They could offer technological help (e.g., installation), offer a platform to be used for free for x months, use computers for families, etc.. This has since been extended by the government, making it a bit bigger, but still, there is a huge need for much better coordination and more support. There is a community where teachers can offer their free time to help families who need it.*

Some actions at the regional level to incentivise the purchase of devices have also been implemented in Italy. Even Estonia, with its high level of digitalisation, has encountered some difficulties. Indeed, during the forum discussion, it was underlined that:

Even in Estonia, the networks are not so complete. Our kids had some issues with the Internet and also with devices. It was nice to see that IT communities started to offer the computers for those families who had problems with enabling the devices for learning, and they donated devices for free!

Moreover, there are differences between rural and urban areas. In rural areas, in particular, there are several obstacles to accessing computers and laptops across Europe:

The acute challenges are electricity and technology. Let alone the developed countries, many villages in developing countries don't get proper electricity. Regardless of that, there are so many who use their phones to do their homework. It's possible of course but in the long term, it is going to have serious detrimental effects on the education quality and level.

One problem observed in all countries (albeit to different extents) was insufficient bandwidth, producing delays or connection failures during lessons and video conferences. In fact, not all geographical areas are reached by a broadband connection. This means that in some cases there is a structural gap that represents an obstacle for people connection. This problem was also said to occur in Estonia, where digital tools are part of everyday learning and e-learning days are part of curricula. The digital learning environments created were not designed for such intensive use as in the pandemic crisis, resulting in collapse in the first few days when all schools tried to run them:

In London and the UK, the Wifi has been down a lot! According to the UK providers, 'the population across the UK is using the Wifi connection more than ever'! I have been, personally, 'ICT challenged' from the start of the lockdown! I have been using three different Wifi connections and mobile data from my iPhone!

Therefore, first of all, it is necessary to overcome problems related to connections, considering the implementation of 5G technologies. The large-scale testing of 5G would allow a more efficient connection and therefore, an improvement in online performance and the types of technologies that can be used at distance. This emergency will provide a boost in this direction:

The problem with the Internet connection definitely will change with the development of 5G: bigger data amounts can be transferred and in situations like with the coronavirus it may have benefits.

This is a technological issue, but also a challenge connected with governance and policies related to the adoption of 5G in Europe and worldwide. The lessons learned in other countries can facilitate new reflections about the best technologies and approaches to use in the future. In some countries like Croatia and Serbia, for example, the government provided online learning classes broadcast via TV for primary school students, with online/distance learning starting the day after the lockdown [34]. This allowed all families to be reached and learning to recommence immediately. This is a possible approach that requires a high level of central coordination. Such a positive experience of using TV is very important because it enables us to reflect on the use of a medium that is widely available and accessible to families, particularly in rural areas. Historically, TV has played a significant role in distance learning; for example, in Italy from the post-war period until the 1960s, there was a television programme that offered a literacy degree for families, especially in rural areas where illiteracy rates were high. These examples suggest the use of a multiplatform approach where TV can play a role among the various technologies proposed.

Other issues underlined by the experts were ethics, privacy and copyright related to the intensive use of online devices by learners and to performing online evaluations due to fraud detection:

There is an overly extended exposure of students on digital devices from PCs[1] to tablets to smartphones. These changes have been forced on students and society in less than two weeks, and many ethical steps

---

[1] Personal computers.

have been forgotten for the sake of health. If a change is not opportunely prpared or planned, it can never be sustainable.

In addition, the phenomena of using resources without creators' permission, certification and proper supervision were emphasised. Consequently, actions that increase awareness among users and safe infrastructure for training courses for children were said to be necessary:

I think there is so much more space for hacking/cybercrime. Especially when it comes to children. More of their private information is being stored online. A lot of institutions also use niche services. Especially those institutions that didn't have any remote learning technologies before are most probably not aiming at implementing their own long-term expensive option and would prefer to opt for a short-term solution.

4.2. Pedagogical Challenges

There is not only innovation linked to technological aspects but also the emergence of new pedagogical aspects. Online learning implies revising the approaches used in face-to-face lessons. Experiences of social distancing during the pandemic have enabled us to understand that:

Pedagogical patterns must be different in virtual classrooms. In the virtual classroom, the educator is more like a moderator and consultant, and lessons cannot be arranged as in a physical classroom. Therefore, learning, especially guidance and feedback, should be given in a different way.

Innovations in teaching methods are therefore needed to engage students, stimulating their proactive behaviour, which is difficult to obtain when one is only connected online. In particular, new approaches to maintain children's attention and participation on a screen for a long time are needed.

First of all, in order to plan an adequate pedagogical course for remote teaching, it is necessary to increase the technological skills of all the actors involved. In various countries, challenges related to gaps in digital literacy in education among teachers, students and parents were said to have emerged.

Teachers should be trained to increase digital and other specific skills for online education in order to adequately plan and implement an innovative pedagogical programme. Although students are usually very familiar with the use of digital devices, they may not be prepared to receive remote teaching and it is quite difficult to capture their attention. Furthermore, parents may not have the necessary educational level and language competence in terms of digital skills. Indeed, the forum discussion underlined that:

Criticalities and limitations appeared to be the teachers' skills; students are more skilled in digital issues as they spend a lot of time engaged in digital communication. Teachers need to manage several operational environments and in the beginning it is messy, with technical problems, a lack of knowledge of the options in certain environments, etc.

Countries like Hungary where the day starts with the teacher collecting the digital devices from the children and are highlighted as forbidden tools ... IT education meant Microsoft Office basic use ... these children, parents and teachers now face a huge challenge and it is not a wonder that the change is big, but we are still far from digital education and even online education.

In many countries, primary schools in particular have never widely experimented with online learning, especially in an emergency situation. Nevertheless, the huge amount of content and information available on the web, frequently structured and planned content for primary schools, are limited; teachers have endeavoured to organise and provide structured content, and parents were complaining that they did not know how to help kids with homework. During this pandemic lock-down, when children were asked to get connected for their lessons, parents were frequently involved in smart working.

Moreover, due to the lack of proper digital devices, some students were forced to use a smartphone to watch lessons without optimised digital content. Although mobile learning offers the possibility of ubiquitous computing, there are many technological limitations related to the inferior functionality involved compared to desktop computers [35]. It is also necessary to deal with the issue of optimising the learning of digital content for mobile devices. Optimising content allows reducing the time spent using smartphones, which represents a critical issue for students' levels of attention and concentration.

In this case, IT infrastructure is playing a fundamental role. I am not talking about the lack of proper digital devices. You cannot ask a boy or girl to spend six to eight hours a day watching lessons on a smartphone. This is mainly also due to the lack of optimised digital content. I think the opportunity, in the medium to long term, is that we understand how to be connected, what to say, what to ask students.

An open question is the use of online learning for young children in kindergarten. For example, in Italy, kindergarten and preschool teachers shared children's songs with parents, short educational videos with simple games and readings told by the teachers to maintain a psychological and pedagogical contact with children. There is a wide debate in the literature related to the types of consequences of children's use of ICT devices before they reach school age. Frequently, families also have different attitudes towards technology and online learning. During the forum discussion it was observed that:

Parents who were by now against digital devices and encouraged their children not to use them, even for learning purposes, faced that their children didn't want to use these devices, as they didn't actually know how to use them. This caused in many families a huge issue. I believe that ICT technologies and devices need to be used by children even in their early ages, but there is a need for control over what and how they use them.

Many universities had already started to introduce experiences of online learning to reduce costs. Therefore, they have encountered fewer difficulties relative to other levels of schooling. COVID-19 is proposing an acceleration in this direction. There are many online learning opportunities offered by well-recognised universities and the offer is growing as a consequence of the COVID-19 pandemic. Cambridge University, for example, was the first university in the UK to announce the delivery of all classes fully online up until September 2021. Students at some universities need to pay to attend online classes, while other universities only require payment for examinations. This issue opens a reflection on the classic model of universities and on the development of a hybrid model that enables students to not be fully present, thereby reducing costs for families and making university access more inclusive. During the forum discussion, the need for governments to take action in the very short term to define strategies focusing on digital education assessment was emphasised.

Teaching is different from assessment. This is where remote learning falls short, whether you take biology labs, violin performances, or sports assessments as exams or even graduation projects.

COVID-19 has also given a boost to the open education resources (OER) approach, related to connectivism theory. This initiative enables the collection of large amounts of free educational materials available. For example, the Design Museum in London provides free lesson plans.

In Italy, too, museums have made virtual tours available online. This free content could be used during lessons. The use of multiple channels and certified resources made available by different institutions require training on how to access these content.

A large amount of content is not always usable; indeed, this content needs to be organised. This can be facilitated by building communities for sharing such content in an institutionalised way, thereby increasing common knowledge. This can also help teachers and professors to overcome the criticalities and difficulties of using a cooperative approach:

Professors and teachers have been under more pressure, not all of them have lectures ready to be presented online, but it is educative and gives a nice boost to the next generation's learning. I hope the next steps are e-learning facilities in secondary schools as well.

However, some experts were critical of this approach because, from a business point of view, some institutions might start collecting all free materials and charging for them.

The presence of online learning during emergency periods enables students to remain in touch with their teachers and also with other students. However, some key issues are building a sense of community between learners and teachers and producing interactive and engaging lessons where all students also know each other.

According to the experts, a community of learners and teachers can be built by increasing "human" cyber interaction:

Platforms can support as much human interaction as possible (multimedia): teachers should be perceived as 'humans' and not some other teaching bot or an artificial agent available online. Interaction is key, I feel.

Furthermore, students' engagement could be increased via informal social activities, such as games and social chats:

Learners and teachers should support each other, but teachers may lose their position in learning: learning will go in the direction of cooperation.

The older generations need to catch up with the younger generations' use of IT. Gen Z in general functions in different online communities quite naturally. If there is a joint shared interest it is going to work, I believe.

The engagement problem can also be overcome by using participatory approaches to education in conjunction with the use of online technologies. The use of co-creation platforms in online learning was also suggested, in which students can become more involved, even participating in creating content for lessons.

4.3. Social Challenges

The emergency was said to represent a good opportunity to acquire practices that promote independence and responsibility from the students' side. However, one of the main limitations is the loss of human interaction between teachers and students as well as among students:

Human interaction is fundamental, especially for young students (secondary, primary schools) that need to learn. Only good professors/teachers can do it.

We need face-to-face interactions, we need to feel emotions, and that can not be given by a 100% remote experience.

According to the experts, although the use of ICT "gadgets" is like "an extended arm" for students around the world who feel comfortable with them, there is no substitute for proper teacher-student interaction. To mitigate problems of inclusion, the experts suggested using a blended approach, whenever possible. Blended learning is defined by [36] as "the thoughtful fusion of face-to-face and online learning experiences". It enables perceptions of "human" factors to be intensified and reinforces feelings of community belonging. Certainly, blended learning facilitates interaction, improving collaboration and social relationships among learners and between learners and teachers [37]. In the future, when normal education activities will be able to resume, a balance between learning at school and online learning should be established. Online learning is something that can complement face-to-face lessons. The forum discussion made clear that not all activities can be done online:

Honestly I do not think that everything would go online; I believe that things will go back to 'normal' in the future but there will be a larger percentage of digitally available education. Physical classes will not disappear at all and communities will stay alive; people are social, they need interaction.

A major challenge is to support students with special needs in their learning activities mitigating any risk of inequality and vulnerability. What do we do about schools of children with mental or physical impairments? This question calls for a new pedagogical approach, taking into consideration the potential advantages of technology.

Emergency remote teaching also presents some challenges for parents and teachers. During such an emergency, they may also be working remotely. This produces a problem relating to the availability of ICT devices for all members of a family. If all people are working at home, there is also a problem of physical space where each person can receive a lesson or do her or his work.

The logistics of online learning have to be carefully considered. Indeed, not all families have sufficient rooms to be used by their children:

My eldest studies at an Austrian university and they switched to fully digital when the State decided. It was quite funny seeing the exams when you are sitting at home with all the family, but no one was allowed to be in the same room for five hours. There's still a way to go... there is no copy-paste possible from the regular classroom to the web classroom.

Moreover, remote teaching for children frequently requires parents' presence, which may make it impossible for them to balance their work activities with supporting their children during their online learning experience. Furthermore, some parents do not have adequate literacy to support learning at home:

Small children need their parents' support, so what about the parent who is not able to work because they must play the teacher's role? How do parents work if they also need to look after their children at home? Studying (and playing)? This might end up disrupting the global economic model in the long run: work productivity will go down.

However, in some countries, emergency remote teaching is opening up new opportunities that are not necessarily linked to emergencies. For example in Estonia, parents look at this experience as a good opportunity for implementing online learning at full speed, in particular for families who travel a lot, who work abroad, who have specific needs and so forth.

This last indication is very interesting, as countries such as Estonia are already highly experienced in ICT. Therefore, the core issue is to understand how online learning can be used and integrated with face-to-face learning, additionally considering that many of the challenges are also connected with the need to overcome technological gaps.

5. The Italian Case Study

The experiences of remote teaching highlighted in various countries during this health emergency have some common issues. Here, we analyse the case of Italy. The picture did not return a uniform situation in the country. Indeed, a gap between north and south has emerged, especially considering the availability and types of devices and platforms used. In the northern classes, advanced platforms are being used and over 51% of students regularly attend video lessons. In the southern regions, students are mostly assigned homework to be done and corrected online. The technological endowment of families is the biggest obstacle impeding the definitive affirmation of online learning. Indeed, according to ISTAT [32], 42% of families do not have a PC in southern Italy, while this percentage is 33% in the rest of Italy. About 14.3% of families with at least one minor child do not have a computer or a tablet and in only 22.2% of families does every member have access to at least one PC or tablet. Therefore, it is evident that it is not easy for everyone to access digital learning content. Moreover, in 2019, two out of three 14–17 year olds had low digital skills.

To overcome this digital divide, several institutional or beneficial initiatives have been undertaken during the COVID-19 pandemic. Moreover, in some cities, campaigns to collect and donate technological devices for students who need them have been launched. Furthermore, the government is attempting to provide a practical solution. Indeed, Decree-Law no. 18 of 17 March 2020, no. 18 (Cura Italia) has allocated funds to encourage schools to use e-learning platforms and to equip them with digital tools, or to enhance those already in their possession. Other funds are being addressed to give less well-off students digital devices on free loan and to train school staff. One plan is for schools to receive funds for the purchase of computers and tablets for students who do not have them. Some companies in the ICT sector have even offered free services to support citizens and have collaborated with the Ministry of Education to make completely free platforms for institutions to facilitate the organisation and transmission of teaching. This emergency has made it clear that having widespread diffusion, across all social groups, of technological devices and broadband connections, is fundamental.

In order to have a more complete framework on Italy, we collected some important opinions considering different points of view involving opinion leaders from the world of press, academia and organizations. These sources are significant as they are qualified experts in the learning sector in Italy. They collect and influence the opinions of the public as well as the important choices made by decision-makers with respect to the problem dealt with.

According to Gervasio [38], opinion leader and journalist of a specialized magazine about the school, the biggest advantages of online learning are the overcoming of space-time barriers and increased flexibility in the ways and styles of learning. This enables the customisation of training paths based on the specific skills and objectives to be achieved by each student. However, some obstacles are technological, such as the difficulty of accessing the network, the speed of data transmission, the quality of students' and teachers' ICT skills, the ability to manage time and knowledge of the best ways to interact online with other students (i.e., to manage a feeling of the community). If students do not have the opportunity to access the network on a regular basis, they risk being left behind, inevitably leading to the alienation of some learners, especially if they are not prepared from the outset for a type of collaborative and constructivist learning.

Another important opinion was provided in the interview with Michilli [39], the General Manager of "Fondazione Mondo Digitale", committed to the creation of an inclusive learning society in which innovation, instruction, inclusion and fundamental values are all combined in a holistic vision.

According to her, it is important to capitalise on this crisis by overcoming all the gaps that are emerging. The commitment to innovate the system must come from institutions, with the collaboration of all actors (citizens, enterprises, non-governmental organisations). Italy is one of the countries with the highest rate of mobile phone ownership in the world, but many families do not have a laptop, PC or landline with an Internet connection. It is not easy to use a mobile phone for online learning, as even the most advanced smartphones do not allow adequate interaction for a long time. The infrastructure problem exists for many teachers and families. The Digital Economy and Society (DESI) Index [40] on Italy is clear: the country is far behind both in the use of fast lines and equipment. Other problems are teachers' and students' skills and the teaching approaches used. If on the one hand there is a wide choice of technological platforms, on the other, organised and certified content for learning is very scarce and publishers are still very traditional. Therefore, teachers often use technologies as a support for a traditional didactic form of learning in which frontal instruction prevails. Special attention must be paid to the "specific needs of students with disabilities"; to this end, the Ministry of Education, University and Research (MIUR) has activated "inclusion via the web", a new thematic channel for supporting teachers in online teaching paths addressed to children with disabilities [41]. As regards disabilities, inclusion within the classroom in the past was managed by focusing on collective teaching and group work, while isolation is a further element of exclusion. This is a huge problem that needs special attention and professionals in the sector.

According to Rivoltella (Professor at the "Università Cattolica del Sacro Cuore" on Media Education and Learning Technologies) [42], the emergency is compelling us to rethink teaching practices. He focuses in particular on pedagogical and relational challenges. According to him, it is not enough to put students in front of a computer screen or assign them homework: educational planning is also needed. The biggest teaching challenges are managing students' motivation and attention; it is necessary to create the content and to also give precise indications to the students through the use of synchronous communication (chat and video communication) to interact, clarify doubts and discuss problems. Cooperation between students must be fostered; the real added value of technology is the possibility of sharing, working and cooperating in a group. Technologies can help reinforce feelings of being part of a community and generate new networks of relationships and meanings. They should transform all diversities (disability, language, culture) into a diversity that enriches rather than being an obstacle that adds separation. The quality of the relationship is not a matter of formats or tools and digital is not an alternative to presence. The relationship is the result of educational intentionality and the digital can be one of the ways to ensure it.

At the governance level, it is necessary to take action in the very short term for defining strategies to face and overcome the technological, pedagogical and social challenges discussed before.

In the case of a new emergency, schools need to use online learning. Therefore, the government has indicated that each educational institution should integrate into its annual plan of the educational offer a part related to digital teaching, identifying ways to redesign teaching activities, taking advantage of the lessons learned during the current emergency. The Ministry of Education also has started designing an official online learning platform. Full participation in online learning (where necessary) must be encouraged, whatever students' economic, social and cultural starting point. With this in mind, the government is activating the following actions to face the open challenges:

- protocols with the Professional Order of Psychologists to manage the emotional effects of the lockdown on students, school staff and families;
- agreements with mobile phone companies, for discounts on costs for connection;
- support actions to ensure that local authorities continue to complete the infrastructure that guarantees coverage of the entire national territory with broadband.

Furthermore, MIUR provided some guidelines [43] for the return to school in September 2020, in which the autonomy of the individual educational institutions was highlighted to better organise teaching activities, using forms of flexibility including organising each class into several groups based on students' levels of learning. The guidelines called for less frontal and more laboratory teaching, in small groups and not necessarily in the classroom, but also in different spaces to combine the need for distancing with innovation. The purchase of new furnishings is being favoured, such as new single desks that allow more collaborative teaching. Digital teaching can be integrated with face-to-face teaching but only in a complementary way in secondary school. Agreements between schools and local entities are being strengthened to encourage the provision of other structures or spaces, such as parks, theatres, libraries and cinemas, to carry out further educational activities or alternatives to traditional ones, but ultimately aimed at educational purposes. Furthermore, training courses for teachers and educational staff are being increased, exposing them to innovative teaching and learning methodologies and courses on interdisciplinary teaching models, multimedia technologies, teamwork and the digitalisation of administrative procedures.

6. Conclusions

Emergency online teaching has allowed schools to provide learning largely undisrupted during the school closures forced by the COVID-19 pandemic. However, there are several challenges to be faced. The results of the analysis of the online discussion forum with international experts, the data from ISTAT and statements of opinion leaders in Italy have revealed several technological, pedagogical and social challenges, additionally confirmed by the reference literature.

The technological challenges are mainly related to the unreliability of Internet connections when thousands of students and workers are simultaneously connected as well as the lack of technological devices for many students. This aspect has been underlined by different studies [16–19], particularly in developing countries like Ghana, Malaysia [20–22]. The pedagogical challenges are associated with teachers' and learners' lack of digital skills, the lack of structured content versus the abundance of online resources, learners' lack of interactivity and motivation and the social and cognitive issues that teachers and schools must address in this situation. The lack of interactivity and motivation of students is connected with the social challenge related to the loss of human interaction between teachers and students as well as among students. In order to encourage children's engagement and curiosity, our results suggest the use of more interactive resources to gamify education, in accordance with Thomas and Rogers [18]. Moreover, there are problems related to the lack of physical spaces at home where lessons can be received and, sometimes, a lack of parental support.

Researchers, universities, educational institutions, businesses and policymakers must be involved in providing adequate answers to the challenges emerging from this worldwide experience. Online learning and emergency remote teaching should become a priority for policymakers in different countries, both in Europe and globally. Lessons learned from this emergency enable us to indicate challenges and proposals for action to face these same challenges addressed to policymakers from different countries so that they can address some of the open challenges. Here we reflect on the summary of opinions by experts coming from the online discussion forum and the data analysis of the Italian case study which served to substantiate the following proposals for action to respond to the identified challenges.

1. Reliable network infrastructure needs to be developed. Teachers, students and parents must have connectivity that allows them to be able to take lessons remotely even when other people in the same house are doing other online activities. In fact, the results of the online discussion forum underlined that the intensive use of networks during the pandemic crisis has produced connection failures in several countries, including Estonia, which is technologically advanced. One suggestion of experts was to develop 5G.
2. More affordable devices must be provided. Devices such as tablets or computers to be connected should be less expensive and Governments should give households incentives to buy them. All the involved actors must have suitable devices to follow a lesson remotely in the most comfortable way. This issue was underlined by the experts, in particular for families with more than one child. Moreover, for the Italian case study, the DESI Index shows that many families do not have a laptop or PC, even though this country has the highest rate of ownership of mobile phones in the world. The European Commission (EC) can play a key role in boosting facilities and infrastructure for online learning. This is also in line with the EC action plan to help individuals, educational institutions and education systems to better adapt for life and work in an age of rapid digital change.
3. Diverse modalities (telecourses, TV, radio, online courses) should be used to provide accessible learning experiences for students in remote areas, as already seen in some countries. The experts provided examples of Croatia and Serbia as countries where these modalities have been successfully implemented. This challenge has also been suggested by Eder [26].
4. Systematic training initiatives should be provided to improve teachers' and learners' technological skills in relation to new emerging models and approaches encouraging the effective use of online learning. The results of this study revealed that in various countries there are challenges related to gaps in digital literacy in education among teachers, students and parents. For example, in Hungary, there is no digital education and/or online education.
5. A clear and consistent plan should be developed, providing structured and planned educational material (content, methodologies and common goals) and more adequate e-learning platforms by using interactive suitable digital learning resources (video, animations, quizzes and games) to maintain students' attention. For example, in Italy, there emerged on one hand a wide choice of technological platforms and on the other very poorly organised and certified content for online learning. Co-creation platforms could be developed and made available, encouraging students' participation in content creation and their inclusion in the learning process.

6. Strategies for communication and digital education assessment need to be created. The lack of student feedback has also been underlined by [23]. According to the experts who participated in the forum, teachers should communicate consistently and often with students so that they do not feel isolated and confused. They should maintain constant contact with students, for example by creating a community group, sending them e-mails twice a week and setting up a frequently asked questions (FAQ) section so that all students can benefit from other students' questions. The experts emphasised that a community of learners and teachers can be built by increasing "human" cyber interaction.
7. A blended approach should be used whenever possible to reinforce a feeling of community belonging, thereby improving social interaction and collaboration among learners and between learners and teachers. According to experts, students need face-to-face interactions, so face-to-face lessons should complement online lessons.
8. Technologies that use virtual and augmented reality need to be improved, making them widely accessible and therefore more engaging and inclusive, in order to stimulate students' involvement and interaction. According to experts, some issues include students' online motivation and involvement. The implementation of these new technologies in online teaching could help in this regard.
9. The use of intelligent technologies for remote teaching, like artificial intelligence, needs to be reinforced to encourage personalised, inclusive and participatory online learning paths. This can open up new possibilities and provide added value to online learning, as long as it is integrated with the pedagogical methodologies used by teachers. In fact, in this study a need to personalise learning and make it more effective emerged.
10. More inclusive tools, platforms and devices considering different web content accessibility guidelines (e.g., WCAG 2.0) need to be developed in order to make digital learning resources accessible to a wider range of people with disabilities.

The open challenges emerging from this health emergency may prove crucial in improving the capability to provide effective online learning, in evolving educational models to overcome inequalities and isolation in emergencies and in preventing social exclusion. Policymakers, enterprises, experts, schools, students and families should collaborate closely to develop accessible and smart learning environments, educational resources and tools additionally able to maintain the sociality, inclusiveness and accessibility of education.

This study aimed to collect opinions, information and experiences and to identify challenges at the European level and proposals for action to face these same challenges addressed to the different actors (policymakers, researchers, teachers, etc.) to overcome the problems that arose during the first lockdown related to the COVID-19 pandemic. This study has enabled us to gain a picture during the first crisis of COVID-19 and it does not presume to be exhaustive. We are planning to extend it in the future providing a major empirical and theoretical corroboration to support the list of actions here hypothesized. Moreover, further research will analyse students' perspectives, experiences, attitudes and feelings and compare them across different countries, in order to provide a more comprehensive view of the phenomenon and to attain more detailed results.

Author Contributions: Conceptualisation, F.F., P.G., and T.G.; methodology, F.F., P.G., and T.G.; data analysis, P.G., and T.G.; investigation, F.F., P.G., and T.G.; writing—original draft preparation, F.F. and T.G.; writing—review and editing, F.F., P.G., and T.G.; supervision, F.F.; funding acquisition, F.F., and P.G. All authors have read and agreed to the published version of the manuscript.

Funding: This research was carried out in the framework of the activities of the H2020 project "The HUB for boosting the responsibility and inclusiveness of ICT-enabled research and innovation through constructive interactions with SSH research–HubIT" funded by the European Commission, Grant Agreement No: 769497.

Acknowledgments: We would like to acknowledge the important contribution of all the participants in the forum discussion organised within the framework of the HubIT project (https://www.hubit-project.eu/).

Conflicts of Interest: The authors declare no conflict of interest.

References

1. UNESCO. COVID-19 Educational Disruption and Response. 2020. Available online: https://en.unesco.org/covid19 (accessed on 30 July 2020).
2. UNESCO. Distance Learning Solutions. 2020. Available online: https://en.unesco.org/covid19/educationresponse/solutions (accessed on 7 September 2020).
3. Clark, R.C.; Mayer, R.E. *E-Learning and the Science of Instruction*, 4th ed.; Wiley: Hoboken, NJ, USA, 2016.
4. Nagrale, P. Advantages and Disadvantages of Distance Education. 2013. Available online: https://surejob.in/advantages-anddisadvantages-of-distance-education.html (accessed on 10 September 2020).
5. Brown, C. Advantages and Disadvantages of Distance Learning. 2017. Available online: https://www.eztalks.com/elearning/advantages-and-disadvantages-of-distance-learning.html (accessed on 10 September 2020).
6. Bijeesh, N.A. Advantages and Disadvantages of Distance Learning. 2017. Available online: http://www.indiaeducation.net/online-education/articles/advantages-and-disadvantages-of-distancelearning.html (accessed on 10 September 2020).
7. Eyles, A.; Gibbons, S.; Montebruno, P. Covid-19 school shutdowns: What will they do to our children's education? A CEP Covid-19 analysis Briefing note No. 001. 2020. Available online: http://cep.lse.ac.uk/pubs/download/cepcovid-19-001.pdf (accessed on 10 September 2020).
8. Montacute, R. Social Mobility and COVID-19. 2020. Available online: https://www.suttontrust.com/wpcontent/uploads/2020/04/COVID-19-and-Social-Mobility-1.pdf (accessed on 30 July 2020).
9. Hodges, C.; Moore, S.; Lockee, B.; Trust, T.; Bond, A. The difference between emergency remote teaching and online learning. *Educ. Rev.* **2020**. Available online: https://er.educause.edu/articles/2020/3/the-difference-between-emergency-remote-teaching-and-online-learning (accessed on 10 September 2020).
10. Rudnick, A. Social, psychological, and philosophical reflections on pandemics and beyond. *Societies* **2020**, *10*, 42. [CrossRef]
11. United Nations. Policy Brief: Education during COVID-19 and Beyond. 2020. Available online: https://www.un.org/development/desa/dspd/wp-content/uploads/sites/22/2020/08/sg_policy_brief_covid-19_and_education_august_2020.pdf (accessed on 4 October 2020).
12. UNESCO. Distance Learning Strategies in Response to COVID-19 School Closures. 2020. Available online: https://unesdoc.unesco.org/ark:/48223/pf0000373305 (accessed on 30 July 2020).
13. UNESCO. Education for Sustainable Development Goals: Learning Objectives. 2020. Available online: https://unesdoc.unesco.org/ark:/48223/pf0000247444 (accessed on 4 October 2020).
14. HubIT. Available online: https://www.hubit-project.eu/ (accessed on 7 September 2020).
15. Vlachopoulos, D. COVID-19: Threat or opportunity for online education? *High. Learn. Res. Commun.* **2020**, *10*, 2. [CrossRef]
16. Outhwaite, L. *Inequalities in Resources in the Home Learning Environment (No. 2)*; Centre for Education Policy and Equalising Opportunities, UCL Institute of Education: London, UK, 2020.
17. Bol, T. Inequality in home schooling during the corona crisis in the Netherlands. *First Results LISS Panel* **2020**. [CrossRef]
18. Thomas, M.S.; Rogers, C. Education, the science of learning, and the COVID-19 crisis. *Prospects* **2020**, 1. [CrossRef] [PubMed]
19. Doyle, O. COVID-19: Exacerbating Educational Inequalities? 2020. Available online: http://publicpolicy.ie/papers/covid-19-exacerbating-educational-inequalities/ (accessed on 30 July 2020).
20. Owusu-Fordjour, C.; Koomson, C.K.; Hanson, D. The impact of Covid-19 on learning-the perspective of the Ghanaian student. *Eur. J. Educ. Stud.* **2020**. [CrossRef]
21. Yusuf, B.N. Are we prepared enough? A case study of challenges in online learning in a private higher learning institution during the Covid-19 outbreaks. *Adv. Soc. Sci. Res. J.* **2020**, *7*, 205–212. [CrossRef]
22. Omodan, B.I. The vindication of decoloniality and the reality of COVID-19 as an emergency of unknown in rural universities. *Int. J. Sociol. Educ.* **2020**. [CrossRef]
23. Mukhtar, K.; Javed, K.; Arooj, M.; Sethi, A. Advantages, limitations and recommendations for online learning during COVID-19 pandemic era. *Pak. J. Med. Sci.* **2020**, *36*. [CrossRef] [PubMed]

24. Verawardina, U.; Asnur, L.; Lubis, A.L.; Hendriyani, Y.; Ramadhani, D.; Dewi, I.P.; Sriwahyuni, T. Reviewing online learning facing the Covid-19 outbreak. *J. Talent Dev. Excell.* **2020**, *12*, 385–392.
25. Bernard, R.M.; Abrami, P.C.; Borokhovski, E.; Wade, C.A.; Tamim, R.M.; Surkes, M.A.; Bethel, E.C. A meta-analysis of three types of interaction treatments in distance education. *Rev. Educ. Res.* **2009**, *79*, 1243–1289. [CrossRef]
26. Eder, R.B. The remoteness of remote learning. *J. Interdiscip. Stud. Educ.* **2020**, *9*, 168–171. [CrossRef]
27. UNESCO. COVID-19 Educational Disruption and Response. 2020. Available online: https://en.unesco.org/covid19/educationresponse (accessed on 7 September 2020).
28. Drane, C.; Vernon, L.; O'Shea, S. *The Impact of 'Learning at Home' on the Educational Outcomes of Vulnerable Children in Australia during the COVID-19 Pandemic*; Literature Review Prepared by the National Centre for Student Equity in Higher Education; Curtin University: Bentley, Australia, 2020.
29. Campbell, M.K.; Meier, A.; Carr, C.; Enga, Z.; James, A.S.; Reedy, J.; Zheng, B. Health behavior changes after colon cancer: A comparison of findings from face-to-face and on-line focus groups. *Fam. Community Health* **2001**, *24*, 88–103. [CrossRef] [PubMed]
30. HubIT. Distance Learning and Emergency Remote Teaching: Opportunities and Criticalities at the Time of the Worldwide Health Emergency—SPEAK OUT! Available online: https://www.hubit-project.eu/forum/topic/distance-learning-and-emergency-remote-teaching-opportunities-and-criticalities-at-the-time-of-the-worldwide-health-emergency-speak-out (accessed on 7 September 2020).
31. Braun, V.; Clarke, V. Using thematic analysis in psychology. *Qual. Res. Psychol.* **2006**, *3*, 77–101. [CrossRef]
32. Istituto Nazionale di Statistica (ISTAT). Available online: https://www.istat.it/it/archivio/240949 (accessed on 7 September 2020).
33. European Commission (EC). Responsible Research & Innovation. 2020. Available online: https://ec.europa.eu/programmes/horizon2020/en/h2020-section/responsible-research-innovation (accessed on 10 September 2020).
34. Republic of Croatia, Ministry of Science and Education. Coronavirus—Organisation of distance teaching and learning in Croatia. Available online: https://mzo.gov.hr/news/coronavirusorganisation-of-distance-teaching-and-learning-in-croatia/3634 (accessed on 7 September 2020).
35. D'Andrea, A.; Ferri, F.; Fortunati De Luca, L.; Guzzo, T. Mobile devices to support advanced forms of e-learning. In *Multimodal Human Computer Interaction and Pervasive Services*; Grifoni, P., Ed.; IGI Global: Hershey, PA, USA, 2009; pp. 389–407. [CrossRef]
36. Garrison, D.R.; Vaughan, N.D. *Blended Learning in Higher Education: Framework, Principles, and Guidelines*; Jossey-Bass: San Francisco, CA, USA, 2008. [CrossRef]
37. Guzzo, T.; Grifoni, P.; Ferri, F. Social aspects and Web 2.0 challenges in blended learning. In *Blended Learning Environments for Adults: Evaluations and Frameworks*; Anastasiades, P.S., Ed.; IGI Global: Hershey, PA, USA, 2012; pp. 35–49. [CrossRef]
38. Gervasio, F. Didattica a Distanza, Alcuni Suggerimenti per Svilupparla al Meglio. *Orizzontescuola.it*. 2020. Available online: https://www.orizzontescuola.it/didattica-a-distanza-alcuni-suggerimenti-per-svilupparla-al-meglio/ (accessed on 7 October 2020).
39. Stentella, M. La Scuola e la Sfida Della Didattica a Distanza: Cosa Possiamo Imparare Dall'emergenza Covid-19. *FPA Digital 360*. 2020. Available online: https://www.forumpa.it/temi-verticali/scuola-istruzione-ricerca/la-scuola-e-la-sfida-della-didattica-a-distanza-cosa-possiamo-imparare-dallemergenza-covid-19/ (accessed on 7 September 2020).
40. European Commission (EC). The Digital Economy and Society Index (DESI). 2020. Available online: https://ec.europa.eu/digital-single-market/en/desi (accessed on 7 September 2020).
41. Ministry of Education, University and Research (MIUR). L'inclusione via Web. 2020. Available online: https://www.istruzione.it/coronavirus/didattica-a-distanza_inclusione-via-web.html (accessed on 10 September 2020).

42. Rivoltella, P.C. Scuola. Tecnologia più condivisione: Così si può fare buon e-learning. *Avvenire.it*. 2020. Available online: https://www.avvenire.it/opinioni/pagine/tecnologia-pi-condivisione-cos-si-pu-fare-buon-elearning (accessed on 7 September 2020).
43. Ministry of Education, University and Research (MIUR). Scuola, Presentate le Linee Guida per Settembre. 2020. Available online: https://www.miur.gov.it/web/guest/-/scuola-presentate-le-linee-guida-per-settembre (accessed on 7 September 2020).

Publisher's Note: MDPI stays neutral with regard to jurisdictional claims in published maps and institutional affiliations.

© 2020 by the authors. Licensee MDPI, Basel, Switzerland. This article is an open access article distributed under the terms and conditions of the Creative Commons Attribution (CC BY) license (http://creativecommons.org/licenses/by/4.0/).

MDPI
St. Alban-Anlage 66
4052 Basel
Switzerland
Tel. +41 61 683 77 34
Fax +41 61 302 89 18
www.mdpi.com

Societies Editorial Office
E-mail: societies@mdpi.com
www.mdpi.com/journal/societies

CPSIA information can be obtained
at www.ICGtesting.com
Printed in the USA
BVHW022018050221
599228BV00031B/867